Mathematical Olympiads
for
Elementary School

My First Book of Mathematical Olympiads

(Problem Book)

Educational Collection *Magna-Scientia*

My First Book of Mathematical Olympiads

Mathematical Olympiads *for* Elementary School

(Problem Book)

Michael Angel C. G., Editor

Copyright © 2021 Michael Angel C. G.

Title:
Mathematical Olympiads for Elementary School. My First Book of Mathematical Olympiads. Problem Book.

Edition, Cover and Interior Design:
Michael Angel C. G.

The illustrations and diagrams as well as various parts of the text of the work are English language adaptations based on the Russian language monographs Олимпиада Начальной Школы and Математическая Олимпиада для 5 классов, from 2011 to 2020.

This work is published under a creative commons BY-NC-ND license that allows you to copy and reproduce parts of the work, for non-commercial purposes, provided that the corresponding authorship is granted, a link to the license is provided, and it is indicated if made changes. If you remix, transform, or build on the material, you cannot distribute the modified material. See details of this license at https://creativecommons.org/licenses/by-nc-nd/4.0/

Preface

The Mathematical Olympiads for Elementary School discussed here are essentially the *Mathematical Olympiads for Elementary School* which are open mathematical Olympiads for students from 1st to 4th grade of elementary school, as well as the *Open Mathematical Olympiads of the City for the 5th grade*. Both Olympiads have been held every year in the city of Moscow since 1996 and 2007, respectively; being the Technological University of Russia – MIREA its main headquarters today. Likewise, the first one consists of two rounds, a qualifying round and a final round, both consisting of a written exam. While the second one consists of two independent rounds, one written and one oral. The problems included in this book (Level 1 – 5) correspond to the final round of the first one and to the written round of the second one, held during the years 2011-2020.

This problem book is a collection of 550 math olympiad problems with six levels of difficulty. And it is especially aimed at schoolchildren between 6 and 11 years old, so that the students interested either in preparing for a math competition or simply in practicing entertaining problems to improve their math skills, challenge themselves to solve these interesting problems. This problem book is ideal (and widely recommended) for elementary school children in upper grades or even middle school students, with little or no experience in Math Olympiads and who require comprehensive preparation for any math competition. Likewise, it can also be useful for teachers, parents, and math study circles.

Thus, a total of 550 problems with answers are made available to the students for their comprehensive and rigorous preparation, which are divided into six levels of difficulty 0 – 5, where each level of difficulty 1 – 5 includes problems corresponding to their respective school grade, while level 0 includes a set of adaptation problems for beginners in math olympiads. The students without experience in Math Olympiads are encouraged to start from the Level 0, regardless of their current school grade. In addition, ten exams are included for each level of difficulty, where each exam consists of 8 problems except those of the Level 5 whose exams consist of 15 problems.

To be able to face these problems successfully, no greater knowledge is required than that covered in the school curriculum; however, many of these problems require an ingenious approach to be tackled successfully. Students are encouraged to keep trying to solve each problem as a personal challenge, as many times as necessary; and to parents who continue to support their children in their disciplined preparation. Once an answer is obtained, it can be checked against the answers given after each group of exams.

Sincerely,

The editor

Contents

Preface 5

Problems: Level 0 11
   Exam 1 13
   Exam 2 17
   Exam 3 21
   Exam 4 27
   Exam 5 31
   Exam 6 33
   Exam 7 37
   Exam 8 43
   Exam 9 45
   Exam 10 51

Answers: Level 0 57

Problems: Level 1 65
   Exam 1 67
   Exam 2 71
   Exam 3 75
   Exam 4 79
   Exam 5 83

    Exam 6 87

    Exam 7 91

    Exam 8 95

    Exam 9 99

    Exam 10 103

Answers: Level 1 107

Problems: Level 2 115

    Exam 1 117

    Exam 2 121

    Exam 3 125

    Exam 4 129

    Exam 5 133

    Exam 6 137

    Exam 7 141

    Exam 8 145

    Exam 9 149

    Exam 10 153

Answers: Level 2 157

Problems: Level 3 165

    Exam 1 167

    Exam 2 171

    Exam 3 175

    Exam 4 179

    Exam 5 183

- Exam 6 187
- Exam 7 191
- Exam 8 195
- Exam 9 199
- Exam 10 203

Answers: Level 3 207

Problems: Level 4 215

- Exam 1 217
- Exam 2 221
- Exam 3 225
- Exam 4 229
- Exam 5 233
- Exam 6 237
- Exam 7 241
- Exam 8 245
- Exam 9 249
- Exam 10 253

Answers: Level 4 257

Problems: Level 5 265

- Exam 1 267
- Exam 2 275
- Exam 3 283
- Exam 4 291
- Exam 5 299

Exam 6 307

Exam 7 315

Exam 8 323

Exam 9 331

Exam 10 339

Answers: Level 5 347

Problems: Level 0

Problems – Level 0

Exam 1

Problem 1. How many more cubes does Carl have than Bob?

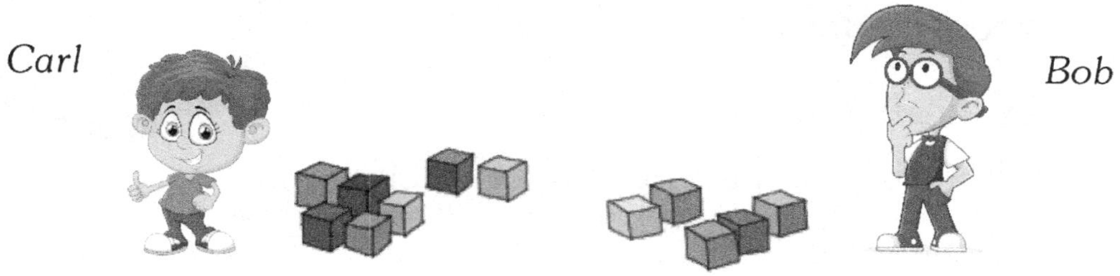

Problem 2. In what figure did Jack draw Sonya correctly without losing any details?

(A) (B) (C) (D) (E)

Problem 3. Bob drew a castle of triangular and quadrangular shapes, shown in the figure. How many quadrangular shapes did he use?

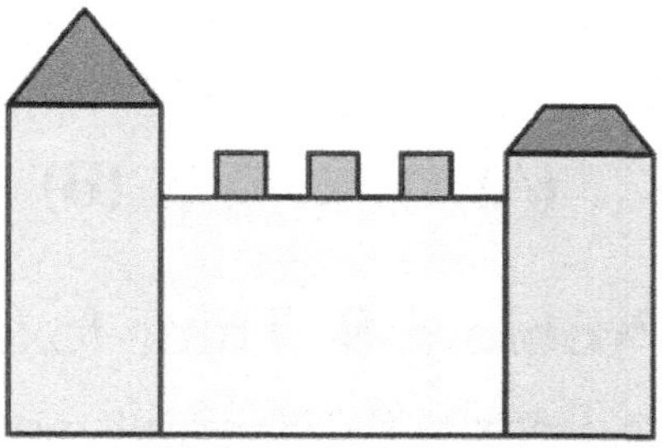

Problem 4. What figure will be obtained if black is replaced by white and white by black in the figure on the right?

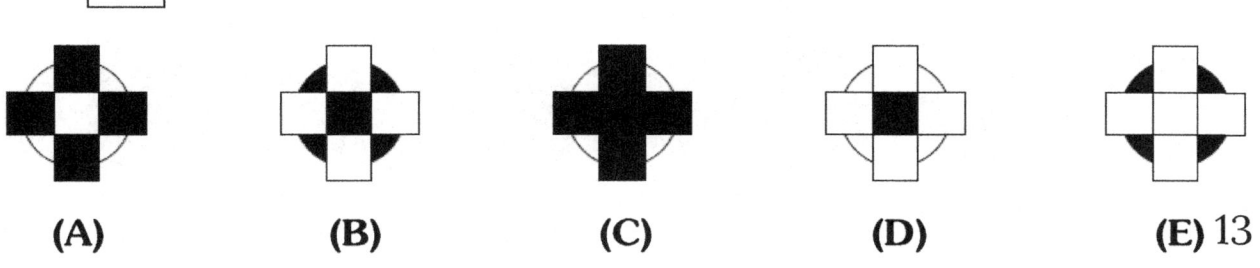

(A) (B) (C) (D) (E)

13

Problem 5. A grandmother divided a cherry pie into as many pieces as grandchildren. If there were 3 cherries on each piece of cake. How many grandchildren does the grandmother have?

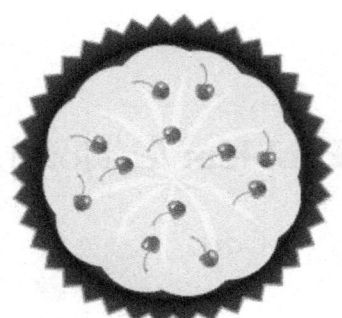

Problem 6. In the figure, each number must be represented so that it is equal to the sum of the two numbers below it.

For example, 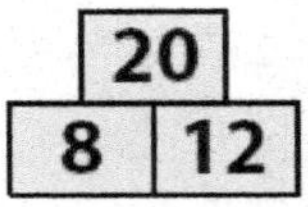 . What number should be in the cell containing the symbol ∗?

Problem 7. In the figure, Luke colors all the squares, where the result is number 13. What pattern will he get?

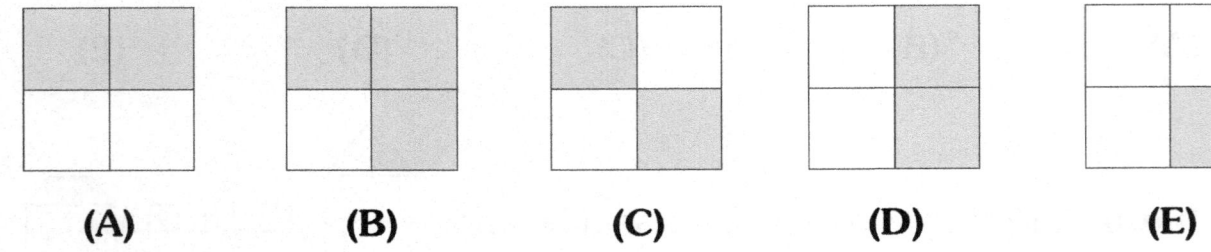

(A) (B) (C) (D) (E)

Problem 8. Three foxes can carry the same load as a zebra. Two zebras can carry the same load as a giraffe. How many foxes does it take to carry the same load that a giraffe and a zebra can carry?

ANSWER SHEET – EXAM 1

Problem 1	Answer

Problem 2	Answer

Problem 3	Answer

Problem 4	Answer

Mathematical Olympiads for Elementary School – *Problem Book*

Problem 5	**Answer**

Problem 6	**Answer**

Problem 7	**Answer**

Problem 8	**Answer**

Problems – Level 0

Exam 2

Problem 1. How many jumps does Jack need to do to reach his mother?

Problem 2. Which of the kangaroos is the tallest?

Joe   Nick   Ralp   Will   Jack

Problem 3. By correctly matching the puzzles shown in the figure, Johnny got three numbers that  satisfy the addition. What number did he get as a result of this addition?

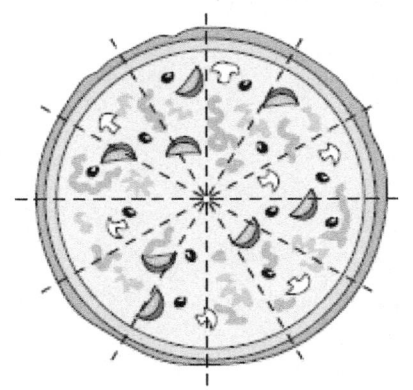

Problem 4. The circle-shaped pizza was cut into 12 pieces. Chris ate two pieces, Joe ate one more piece than Chris, and Bob ate three more than Chris. How many pieces were left if no one else ate pizza?

Problem 5. A dad has to change the wheels of five tricycles. Two wheels on the first and last tricycle, and three on all others. How many wheels does dad have to replace?

Problem 6. Lisa makes the numbers from the matches as shown in the figure below:

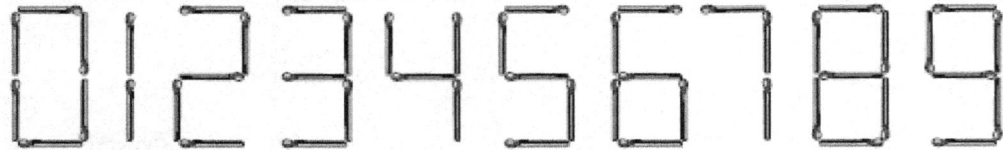

How many matches will she need to make the number 2018 according to this model?

Problem 7. Roxane baked some cakes. She ate a cake, then with Olenka they ate one each, then with Olenka and Vanessa they ate one each, and later with Olenka, Vanessa and her mother they also ate one each. If one cake left. How many cakes did Roxane bake?

Problem 8. There are several bees in a hive. 13 of them flew to collect pollen, but only 4 returned to the hive for lunch. There are now 21 bees in the hive. How many bees were in the hive at the beginning?

Problems – Level 0

ANSWER SHEET – EXAM 2

Problem 1	Answer
Problem 2	**Answer**
Problem 3	**Answer**
Problem 4	**Answer**

Mathematical Olympiads for Elementary School - *Problem Book*

Problem 5	Answer

Problem 6	Answer

Problem 7	Answer

Problem 8	Answer

Exam 3

Problem 1. What frame shape is not used in the pictures?

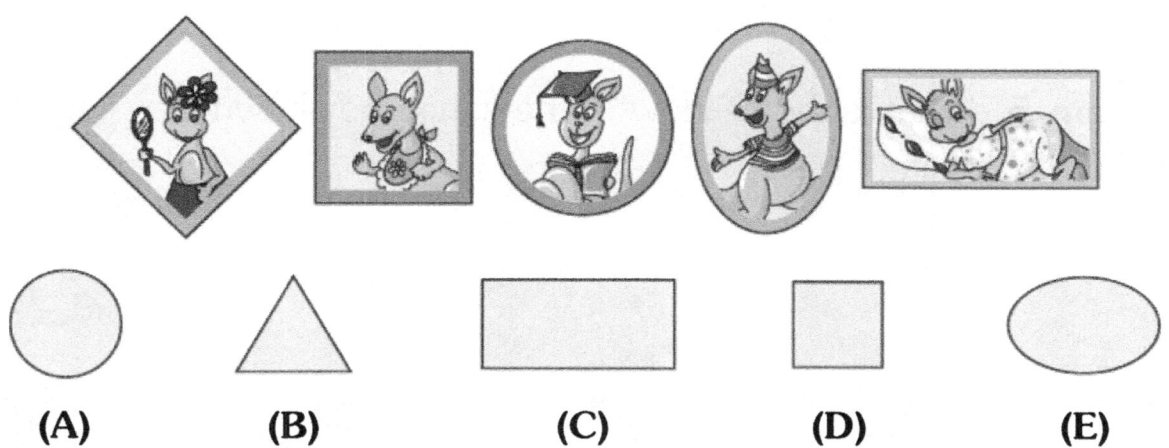

Problem 2. In a tree there were three crows, three cats, two titmice and Ivan. How many birds were on the tree?

Problem 3. Which of the options shows the correct weight of the basket of apples?

(A) 13 + 4 (B) 18 - 4 (C) 11 + 5

(D) 12 + 3 (E) 17 - 5

Problem 4. Jack and Robin represent different numbers in the equality shown. If Robin represents the greatest number. What number is Jack representing?

Mathematical Olympiads for Elementary School – *Problem Book*

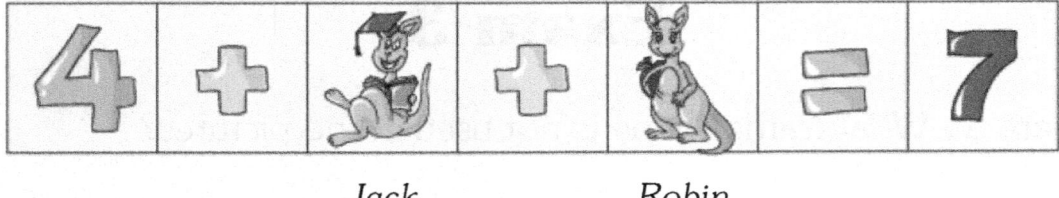

          Jack   *Robin*

Problem 5. What is the shadow of the kangaroo shown in the figure opposite?

(A)   (B)   (C)   (D)   (E)

Problem 6. Fifteen children participate in a dance group, of which one is a boy and fourteen are girls. How many girls will participate in total after Sarah, Luke and Zoe have signed up for this group?

Problem 7. From which figure shown in the options can the house shown in the figure to the side be made?

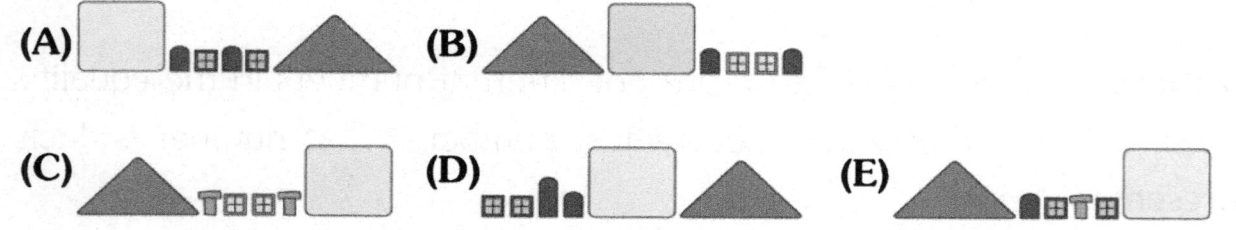

Problem 8. There are two stories in a book. The first story occupies 11 pages of text and 3 pages of images. While in the second - 12 pages of text and 4 pages of images. How many total pages do the two stories take up?

ANSWER SHEET – EXAM 3

Problem 1	Answer
Problem 2	**Answer**
Problem 3	**Answer**
Problem 4	**Answer**

Problem 5	Answer
Problem 6	**Answer**
Problem 7	**Answer**
Problem 8	**Answer**

Exam 4

Problem 1. Ralph has a flag with the number 10. Will wanted to compete with Ralph in math and wrote on some flags the sum of the numbers that he thought were equal to 10. Which flag is wrong?

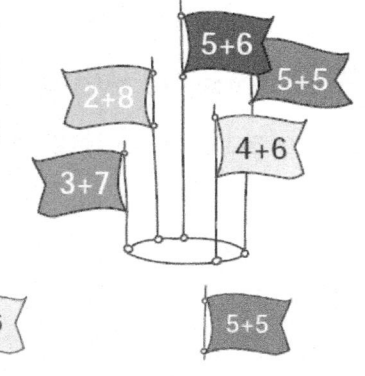

(A) 3+7 (B) 2+8 (C) 5+6 (D) 4+6 (E) 5+5

Problem 2. The spider can move down, to the right, or to the left, but not up. If it must return at every intersection. What is the closest exit to the spider?

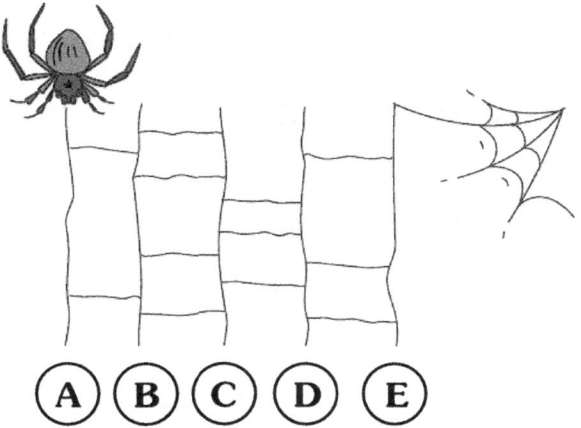

Problem 3. Help Kangaroo Jack to order the numbered cards from the largest to the smallest. Which is the right answer?

(A) 55 29 34 17 2
(B) 34 29 17 2 55
(C) 34 55 29 17 2
(D) 29 34 17 55 2
(E) 55 34 29 17 2

Mathematical Olympiads for Elementary School – *Problem Book*

Problem 4. Two years ago, the sum of the ages of Bob and Rob was 15 years. Now Bob is 10 years old. In how many years will Rob be 11 years old?

Problem 5. What is the fewest matches that need to be added to the figure beside to get a square?

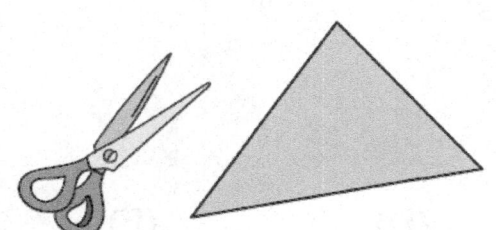

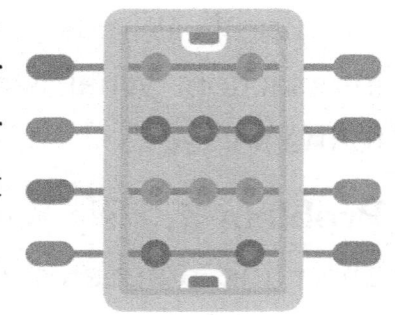

Problem 6. From the triangle shown in the figure to the side, cut 3 corners. How many angles does the new figure have?

Problem 7. Luke, Johnny and Bobby ate apples. Johnny ate 5 apples, Bobby - 3 apples. Luke and Johnny together ate three times the amount Bobby ate. How many apples did Luke eat?

Problem 8. Matthew and Joseph played foosball. The first match ended with a score of 2: 0 in favor of Matthew, the second match ended 2: 4 in favor of Joseph. Who won both games and with what score?

Problems – Level 0

ANSWER SHEET – EXAM 4

Problem 1	Answer
Problem 2	Answer
Problem 3	Answer
Problem 4	Answer

Problem 5	Answer
Problem 6	**Answer**
Problem 7	**Answer**
Problem 8	**Answer**

Problems – Level 0

Exam 5

Problem 1. Each student told the teacher how many fairy tales they knew. The teacher wrote it on the board (see figure).

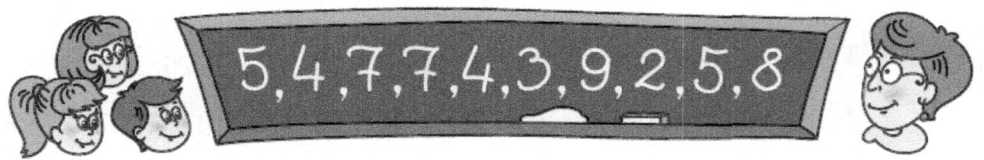

How many students knew less than 6 fairy tales?

Problem 2. From a 60 cm long loaf of bread, my mother cut a 15 cm long piece from each end for my twin brothers.

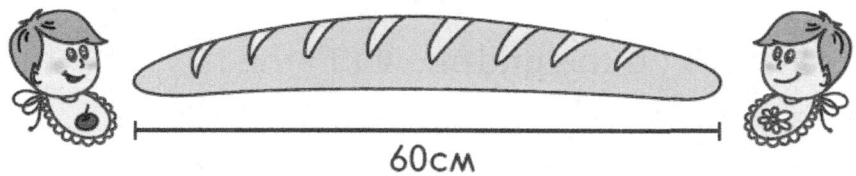

What is the length of the remaining part?

Problem 3. Tanya and Julia each received a bar of chocolate. Tanya has a whole bar and Julia ate part of it (see figure). How many squares of chocolate did Julia eat?

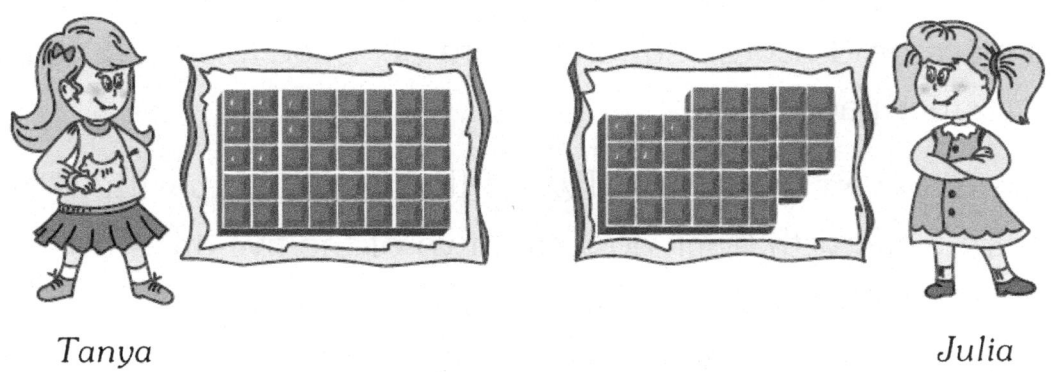

Tanya   Julia

Mathematical Olympiads for Elementary School – Problem Book

Problem 4. There are 12 balls, 10 dice, and 7 small boxes in a large box. There is one candy in each small box. How many items (balls, cubes, boxes, and candies) are in two large boxes?

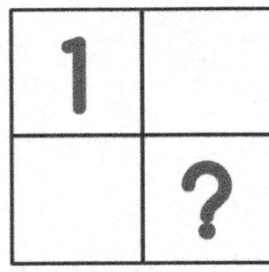

Problem 5. Nick wrote the numbers 1, 2, 3 and 4 in the cells of the 2 × 2 table (each number only once). He wrote the number 1 as shown in the figure. The sum of the numbers in the cells that have a common side with the cell in which the number 3 is written is equal to 5. What number did Nick write in the cell indicated by the question mark?

Problem 6. The letters K A N G A R O O are placed at some nodes on the grid shown in the figure. The side of the small square on this grid is 1 m. Go from letter P to letter Q along the grid lines and collect these letters in the specified order. What is the shortest length of that path?

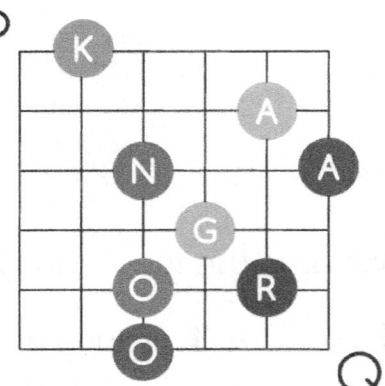

Problem 7. Jack places his photos in an album. If he places a photo on a page, he uses 4 stickers. If he places two photos on one page, he uses 6 stickers. How many stickers does Jack need to place on 3 pages with one photo per page and on 3 pages with two photos per page?

Problem 8. Albert has 5 dice. Two of the five are gray with white dots, the rest are white with black dots, and two of the five are small, the rest are large. Which figure shows Albert's cubes?

(A)  (B)  (C)

(D)  (E)

ANSWER SHEET – EXAM 5

Problem 1	Answer
Problem 2	**Answer**
Problem 3	**Answer**
Problem 4	**Answer**

Problem 5	Answer
Problem 6	Answer
Problem 7	Answer
Problem 8	Answer

Problems – Level 0

Exam 6

Problem 1. A mom made all-digit cookies for her children. Her youngest son Bob ate some cookies. The remaining cookies are shown in the figure opposite. Which cookies will Bob's siblings no longer be able to eat?

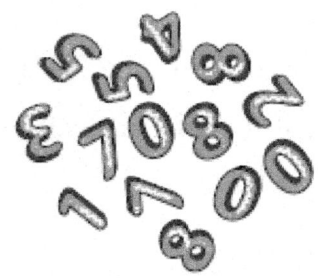

Problem 2. During the night, 3 kangaroos, 2 lion cubs, 1 beaver and 4 monkeys were born in the zoo. How many animals were born that night at the zoo?

Problem 3. Rick places two-colored balls on the steps of a staircase so that for each next step, he adds a ball, whose color differs from the color of the previous ball, that is, as shown in the figure. How will the balls be placed on the step marked with a question mark?

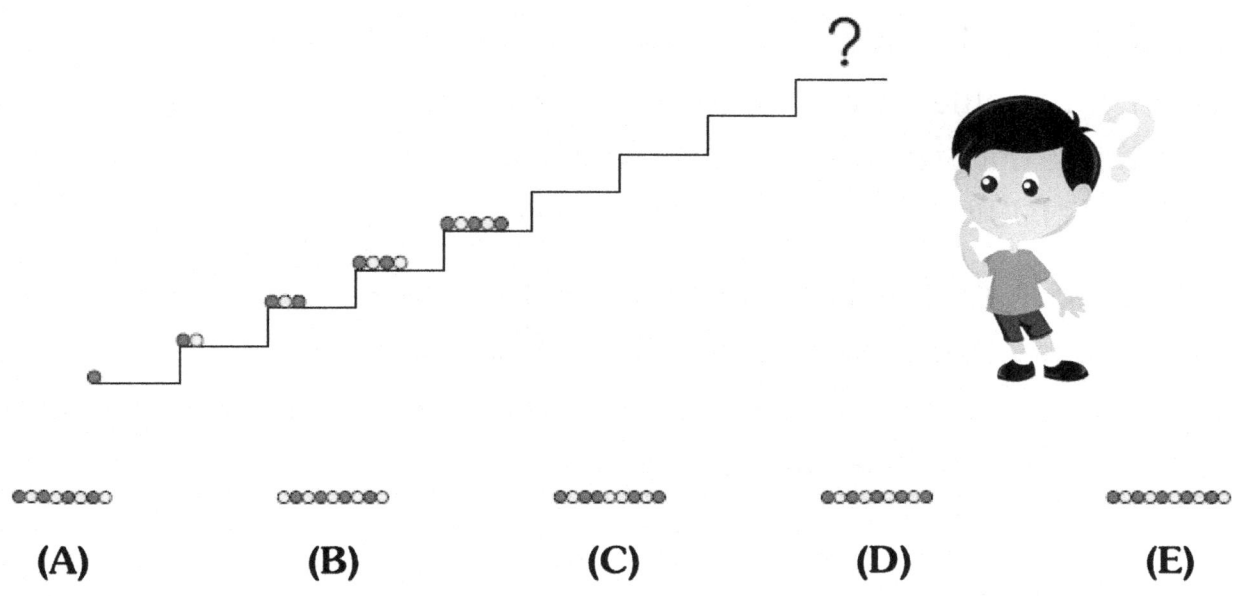

(A)   (B)   (C)   (D)   (E)

Problem 4. A group of 7 girls and 9 boys bought theater tickets in advance. Of these, 2 girls and 1 boy became ill and could not go to the theater. How many children in this group were able to attend the theater?

Problem 5. Only 5 digits were used to write the numbers 0, 1, 11, 12, 9, 8, and 19. What is the minimum number of numbers that need to be removed so that only three digits are used to write the remaining numbers?

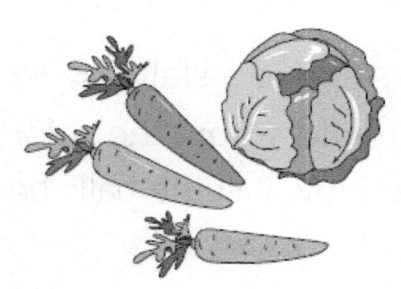

Problem 6. My rabbit eats three carrots or a whole cabbage a day. It ate 12 carrots in a certain week. How many whole cabbages did my rabbit eat in that week?

Problem 7. A bottle of milk costs 12 coins. The milk in this bottle costs 10 coins more than an empty bottle. How much does an empty bottle cost? (1 coin <> 100 cents)

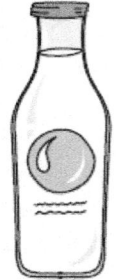

Problem 8. One day three children met. Everyone greeted each other, and each greeting consisted of just two different words. How many words did these kangaroos say if it is known that no word was repeated twice?

Problems – Level 0

ANSWER SHEET – EXAM 6

Problem 1	Answer
Problem 2	**Answer**
Problem 3	**Answer**
Problem 4	**Answer**

Problem 5	Answer
Problem 6	Answer
Problem 7	Answer
Problem 8	Answer

Problems – Level 0

Exam 7

Problem 1. A student solved five exercises, but made a mistake. Help the student discover in which exercise he was wrong.

(A) $20 + 1 - 7 = 14$ (B) $20 - 1 + 7 = 26$ (C) $2 + 0 + 17 = 19$

(D) $2 - 0 + 17 = 18$ (E) $20 - 1 - 7 = 12$

Problem 2. The figure shows 12 beads. Chris painted the first, third, and seventh bead in blue. The second and fifth in green. He did not paint others. How many beads were left unpainted?

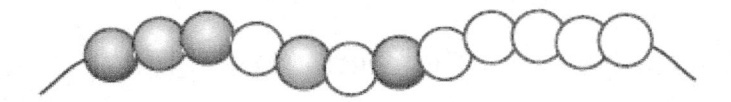

Problem 3. Help Peter to determine how many balls are in the third box, if there are 2 more balls in the second box than in the first and 4 more than in the third.

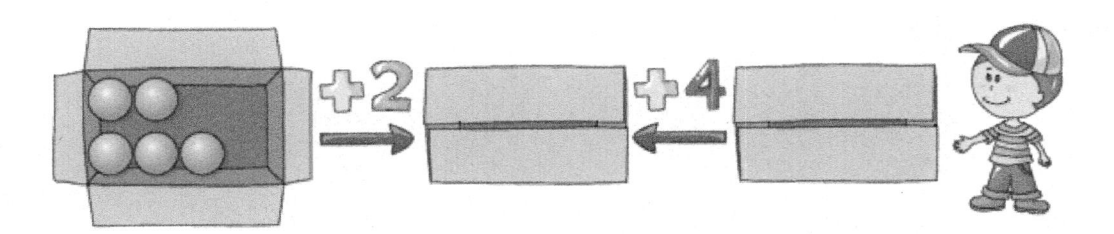

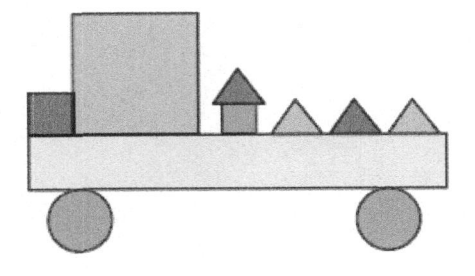

Problem 4. How many triangles and how many squares are shown in the figure?

Problem 5. At the zoo, a rabbit eats 1 *kg* of cabbage, 2 *kg* of carrots and potatoes a day. If he eats potatoes and cabbage as much as carrots and cabbage. How many kilograms of potatoes does the rabbit eat per day?

Mathematical Olympiads for Elementary School – *Problem Book*

Problem 6. Snail and Hedgehog's houses are located in different parts of the city from a fairy tale. The figure shows schematically the different paths between the houses as well as the distances between them in meters. What is the shortest distance between their houses?

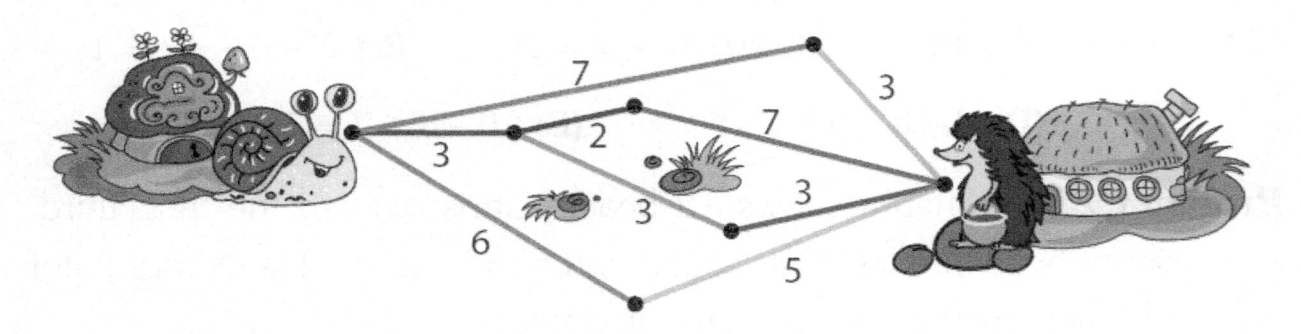

Problem 7. How many light gray squares need to be painted dark gray so that the number of light gray squares equals the number of dark gray squares?

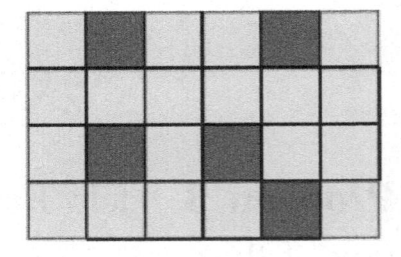

Problem 8. Which of the figures suggested in the options is NOT part of the landscape?

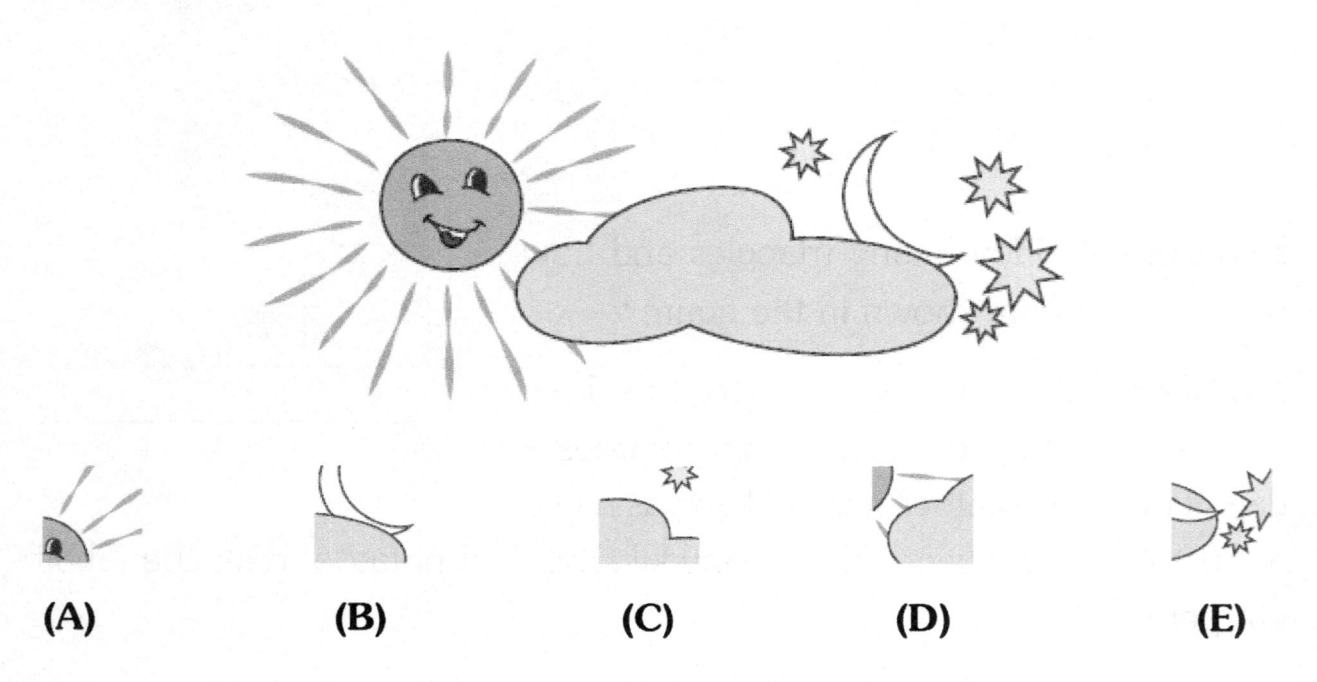

(A)   (B)   (C)   (D)   (E)

Problems – Level 0

ANSWER SHEET – EXAM 7

Problem 1	Answer
Problem 2	**Answer**
Problem 3	**Answer**
Problem 4	**Answer**

Problem 5	**Answer**
Problem 6	**Answer**
Problem 7	**Answer**
Problem 8	**Answer**

Problems – Level 0

Exam 8

Problem 1. The figure shows circles and triangles. How many triangles are completely located (drawn) inside the circles?

Problem 2. How many times is the number 2018 written in the columns and rows of the table shown to the right?

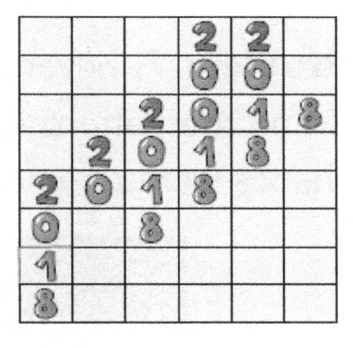

Problem 3. Mathew created a coding system to write the word MATEMATIKA using the numbers given in the table:

M	A	T	E	M	A	T	I	K	A
13	7	8	25	13	7	8	2	6	7

When he coded the word MIKE according to this system, he got the following set of numbers:

(A) 713713   (B) 132627   (C) 137287   (D) 825871   (E) 132625

Problem 4. In the land of adorable cats, a chocolate fish costs two cat coins and a candy mouse costs three cat coins.

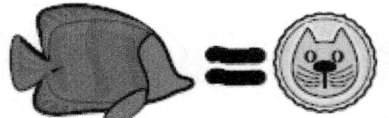

43

Kitty Murzia bought three chocolate fish and a candy mouse. How many coins did she pay?

Problem 5. The figure shows seven paths in the park. Mary always starts her walk at point "O" and only goes through two paths. Which of the points in the park shown in the image could she never reach during the walk?

Problem 6. A cake was cut into 12 pieces. Bob and Matt ate 2 pieces each, Louis and John ate 3 pieces each. How many pieces are left?

Problem 7. Several shapes ◆ were removed from the pattern in Figure 1 and the pattern shown in Figure 2 was obtained. How many of those shapes were removed?

Figure 1

Figure 2

Problem 8. Kangaroo Jack wants to get the carrot by collecting as many coins as possible. He can jump to the next cell in each of the four possible directions, as shown in Figure 2. If the number of coins in each cell is shown in Figure 1. What is the largest number of coins Jack can collect in his journey to the carrot?

Figure 1

Figure 2

ANSWER SHEET – EXAM 8

Problem 1	Answer
Problem 2	**Answer**
Problem 3	**Answer**
Problem 4	**Answer**

Problem 5	Answer

Problem 6	Answer

Problem 7	Answer

Problem 8	Answer

Problems – Level 0

Exam 9

Problem 1. The snowman has five buttons, a pom pom hat, and a long nose. What image shows the snowman?

(A)   (B)   (C)   (D)   (E)

Problem 2. Anna classified the figures into three groups by color; see the figure below.

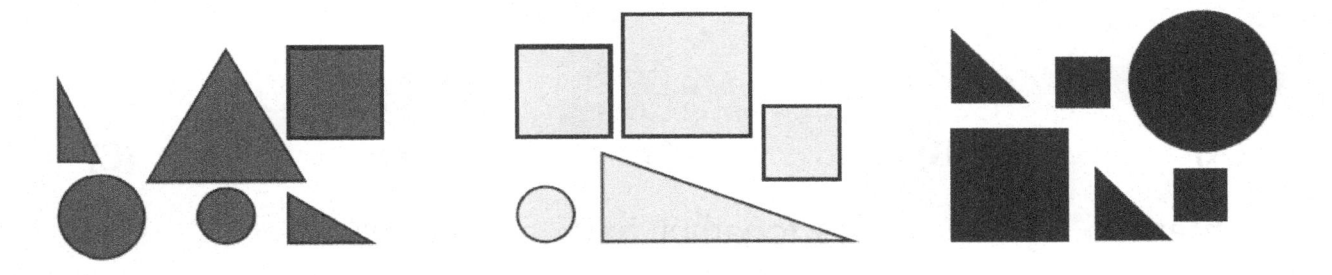

Peter re-classified the same shapes into three groups: triangles, squares, and circles. Which of his groups has the greatest number of shapes?

Problem 3. Albert plays a game, moving a chip as many positions as the die shows. He starts from cell 3 and rolls the die twice. If he first got 4, and then 5. To what cell number will Albert move his chip according to the rules of the game?

Problem 4. All the puppies in the figure weigh the same and the weights are in kilograms.

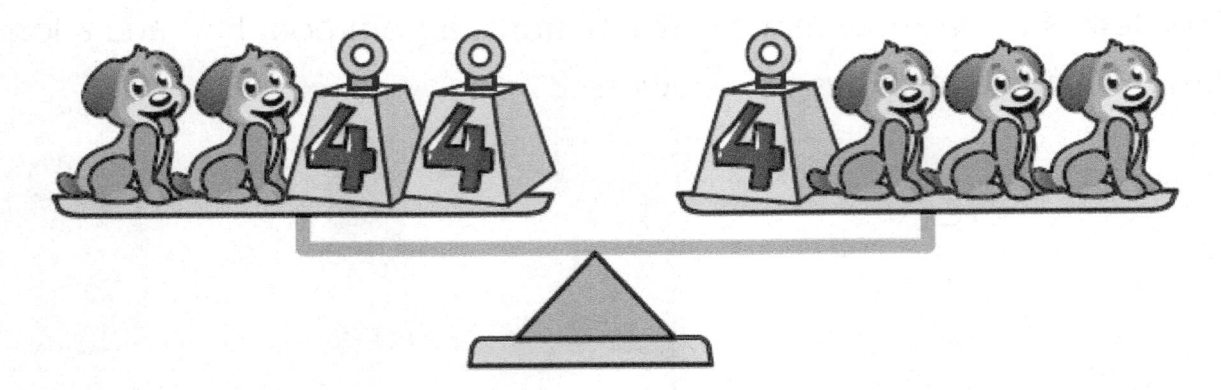

How much does a puppy weigh if the scale in the figure is balanced?

Problem 5. Sarah makes a necklace, stringing beads always in the same order (see figure). In what order should Sarah string four additional beads?

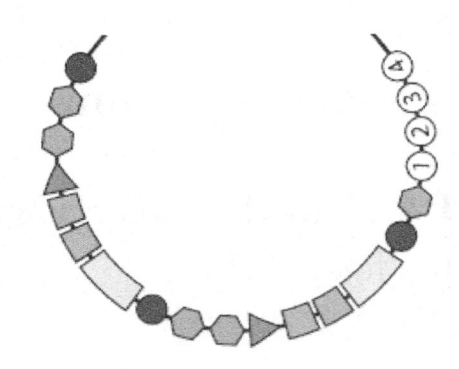

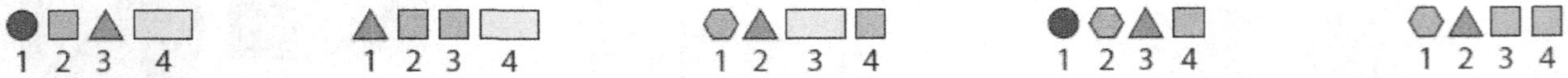

(A) (B) (C) (D) (E)

Problem 6. Three daisies together have exactly 20 petals. Two of them are shown in the figure to the right. In which option can the image of the third daisy be seen?

(A) (B) (C) (D) (E)

Problem 7. Peter made a 9-cube arrangement on the floor. Then, he drew the top view of it. Which of the proposed figures is the arrangement made by Peter?

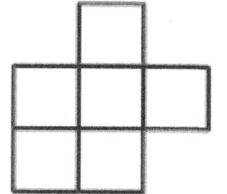

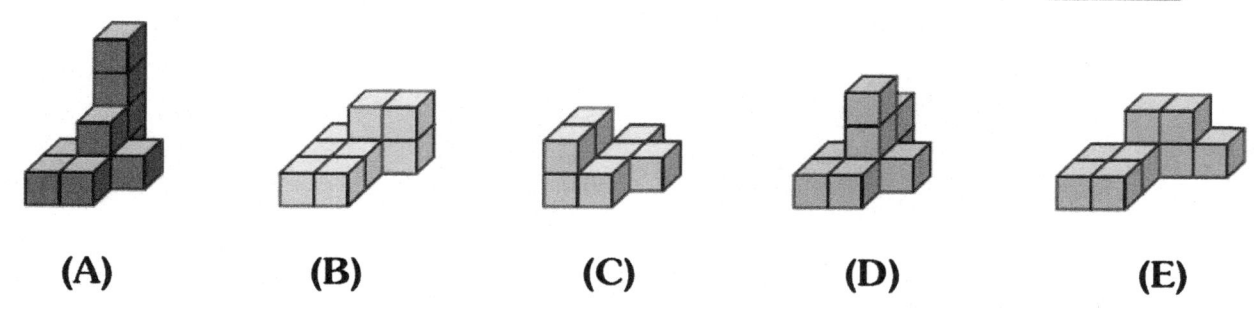

(A) (B) (C) (D) (E)

Problem 8. When you enter Grace's room through the door, you will see that there is a closet on your right and a window on your left. A chair to the right of the window and a shelf to the right of the closet. Which figure schematically shows Grace's room?

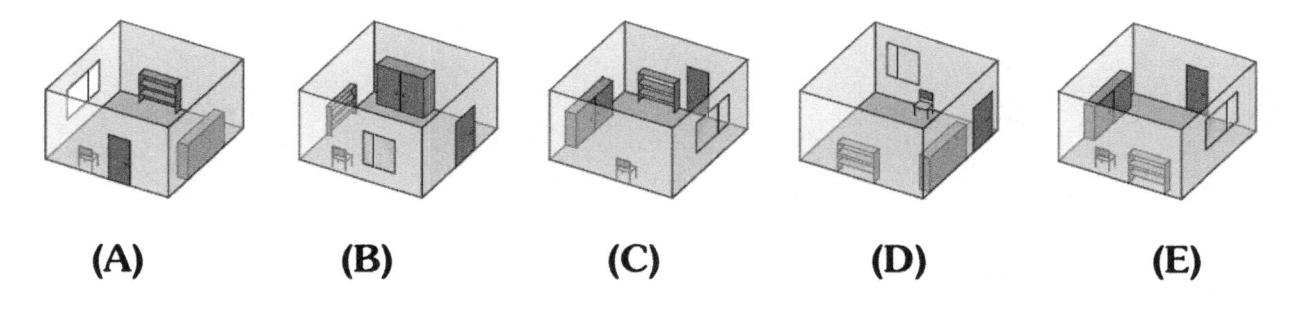

(A) (B) (C) (D) (E)

ANSWER SHEET – EXAM 9

Problem 1	Answer
Problem 2	**Answer**
Problem 3	**Answer**
Problem 4	**Answer**

Problem 5	**Answer**
Problem 6	**Answer**
Problem 7	**Answer**
Problem 8	**Answer**

Problems – Level 0

Exam 10

Problem 1. A box has balls of different colors: 2 red, 1 green, 3 blue, 2 yellow, and 1 purple. John drew 2 red, 1 green, 2 blue, 2 yellow and 1 purple from the box. What color is the ball that was left in the box?

Problem 2. A magician takes toys out of his hat in the same order:

What are the next two toys that the magician will take out from the hat?

(A)   (B)   (C)   (D)   (E)

Problem 3. The figure shows several pencils of the same size placed on the teacher's desk. How many pencils do not touch the desk?

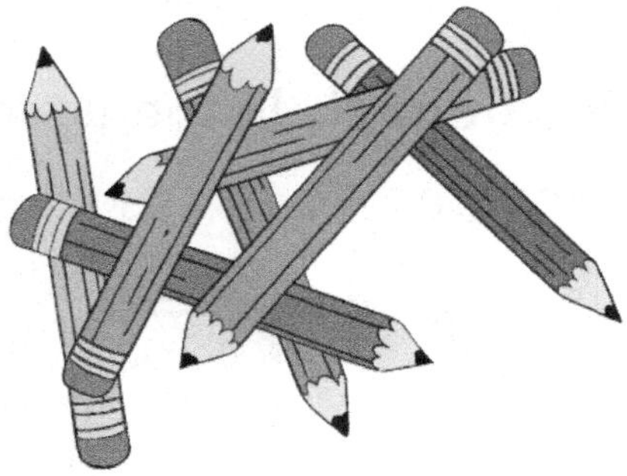

Problem 4. A beetle goes through a garland with flags. It started at the end marked with ∗ and is now between the two adjacent flags shown in the figure beside. How many flags will the beetle find on its way when it reaches the end of the garland without returning?

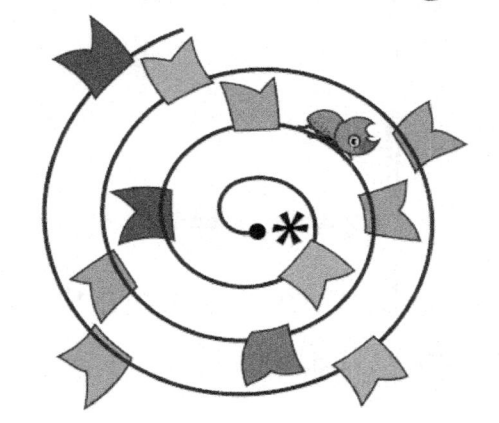

Problem 5. Which of the figures suggested in the options below is not a circle, is not a square and is not colored?

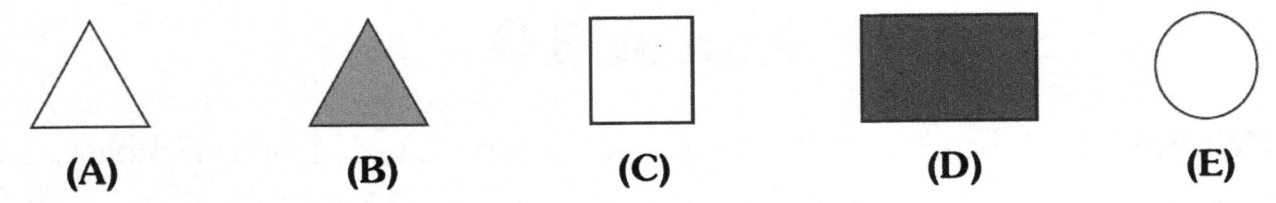

Mathematical Olympiads for Elementary School – *Problem Book*

(A) (B) (C) (D) (E)

Problem 6. How many sheep can be seen on the puzzle of four pieces when they are put together correctly?

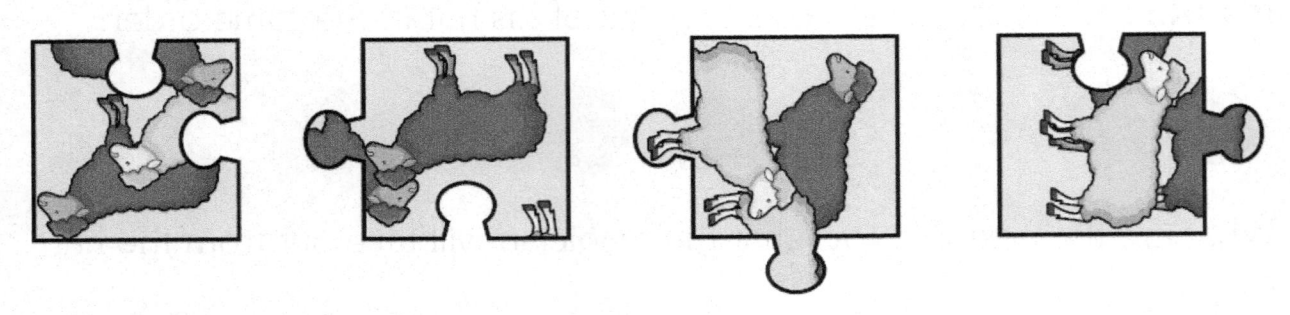

Problem 7. Identify a set that has only numbers greater than 4 and less than 10.

(A) 8,7,5,3   (B) 11,8,6,4   (C) 5,9,7,2   (D) 5,6,7,8   (E) 10,8,5,1

Problem 8. In the figure made of matches shown below, it can be seen 5 squares.

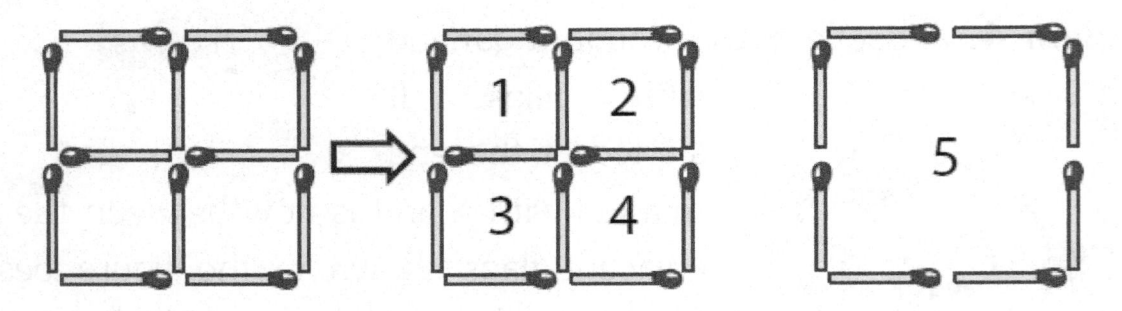

What is the least number of matches that have to be removed from this figure so that no squares can be seen in the resulting figure?

Problems – Level 0

ANSWER SHEET – EXAM 10

Problem 1	Answer
Problem 2	**Answer**
Problem 3	**Answer**
Problem 4	**Answer**

Problem 5	Answer

Problem 6	Answer

Problem 7	Answer

Problem 8	Answer

Answers: Level 0

Answers - Level 0

Exam 1	Exam 2
1. 2 cubes.	**1.** 12 jumps.
2. Option (C).	**2.** Jack.
3. 7 quadrangular shapes.	**3.** 75.
4. Option (A).	**4.** 2 pieces.
5. 4 grandchildren.	**5.** 13 wheels.
6. 5.	**6.** 20 matches.
7. Option (D).	**7.** 11 cakes.
8. 9 foxes.	**8.** 30 bees.

Answers – Level 0

Exam 3	**Exam 4**
1. Option (B).	**1.** Option (C).
2. 5 birds.	**2.** Exit A.
3. Option (D).	**3.** Option (E).
4. 1.	**4.** In 2 years.
5. Option (C).	**5.** 7 matches.
6. 16 girls.	**6.** 6 angles.
7. Option (E).	**7.** 4 apples.
8. 30 pages.	**8.** Tie 4:4.

Answers - Level 0

Exam 5	**Exam 6**
1. 6 students.	**1.** 6 and 9.
2. 30 *cm*.	**2.** 10 animals.
3. 6 squares of chocolate.	**3.** Option (D).
4. 74 items.	**4.** 13.
5. 4.	**5.** 2 numbers.
6. 24 m.	**6.** 3 cabbages.
7. 30 stickers.	**7.** 1 coin.
8. Option (C).	**8.** 12 words.

Answers – Level 0

Exam 7	Exam 8
1. Option (D).	**1.** 2 triangles.
2. 7 beads.	**2.** 6 times.
3. 3 balls.	**3.** Option (E).
4. 4 triangles and 3 squares.	**4.** 9 coins.
5. 2 *kg*.	**5.** Point E.
6. 9 *m*.	**6.** 2 pieces.
7. 7 squares.	**7.** 11 shapes.
8. Option (E).	**8.** 24 coins.

Answers - Level 0

Exam 9	**Exam 10**
1. Option (E).	**1.** Blue.
2. Squares.	**2.** Option (E).
3. 12.	**3.** 4 pencils.
4. 4 *kg*.	**4.** 8 flags.
5. Option (E).	**5.** Option (A).
6. Option (A).	**6.** 8 sheep.
7. Option (D).	**7.** Option (D).
8. Option (C).	**8.** 3 matches.

Answers - Level 0

Problems: Level 1

Problems – Level 1

Exam 1

Problem 1. The hedgehog has a 4-wheel bike, Krosh has a 3-wheel bike, and Nyusha has a 2-wheel bike. Krosh, Nyusha, and the hedgehog went for a walk, some on bikes, some on foot. Losyash counted the number of wheels: it turned out 7. Who went out for a walk on foot?

Problem 2. A portion of a large gift-wrapping paper has been cut out. Petya found 6 different pieces. Which of the pieces belongs to this paper?

Problem 3. Gosha folded square paper in half and then in half again. Then, he pricked the resulting square in the middle with a fork. How many holes will Gosha see when he unfolds the square?

Problem 4. In a 4-cell square, Olya painted more than 3 cells with a yellow pencil, and then Ira painted more than 3 cells with a blue pencil. It turned out that all the cells in the square are painted. Also, if you paint the cell with yellow and blue pencils, you get green. How many green cells are in the square?

Problem 5. Cut the figure shown in the picture into three equal pieces.

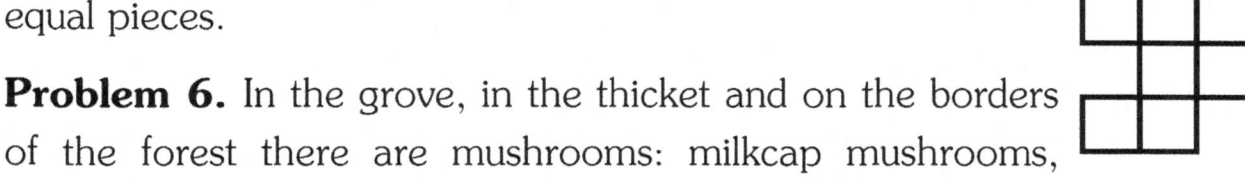

Problem 6. In the grove, in the thicket and on the borders of the forest there are mushrooms: milkcap mushrooms, poplar mushrooms and honey mushrooms. And in each place there is only one species of mushroom. Petya went to gather mushrooms in the thicket and grove, and brought milkcap mushrooms and poplar mushrooms. Vasya went to the grove and the borders of the forest and returned with poplar mushrooms and honey mushrooms. If Dima goes for the milkcap mushrooms. Where should he go?

Problem 7. Winnie the Pooh, Piglet, Donkey Eeyore and Christopher Robin play in a seesaw (see image). It is known that Winnie the Pooh and Piglet together weigh more than Donkey Eeyore, and that Eeyore weighs more than Christopher Robin and Piglet together. Who weighs more if Christopher Robin and Winnie the Pooh sit on the seesaw?

Problem 8. "I love oranges", Sonya said. "No, I love oranges. You love apples", Vera said. "I don't like oranges", Andrei said. It is known that there is at least one child who likes each fruit and that no child likes both fruits at the same time. Which child loves apples and which child loves oranges, if they all lie?

ANSWER SHEET – EXAM 1

Problem 1	Answer
Problem 2	**Answer**
Problem 3	**Answer**
Problem 4	**Answer**

Problem 5	Answer
Problem 6	**Answer**
Problem 7	**Answer**
Problem 8	**Answer**

Problems – Level 1

Exam 2

Problem 1. My sister's name is Anna Pavlovna. My mother's name is Svetlana Dmitrievna and my grandfather's name is Ivan Petrovich. What is my father's name?

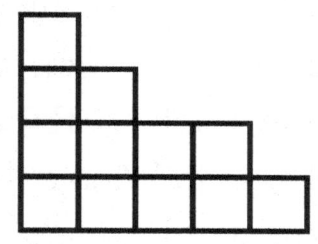

Problem 2. Cut the checkered figure shown below into two equal parts.

Problem 3. Andryusha photographed reflections of fruits in the mirror. And then he lost a picture. Which picture is Andryusha missing now?

Problem 4. In the table beside, arrange the circles, triangles, squares, and crosses so that in each column and each row, as well as each highlighted small square, there are all four figures.

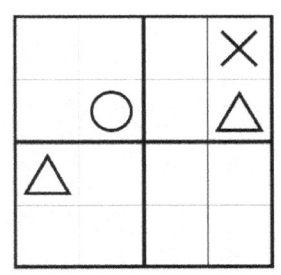

Problem 5. Three cars participating in a race came out in this order: yellow, red, blue. They reached the finish line in this order: "Honda", "Mercedes", "Audi". At the same time, no car finished the same way it started. What color is each car brand, if Audi is not yellow?

Problem 6. Paths have been built in the Enchanted Forest (see image). It turned out that the path from Piglet to Eeyore Donkey is 7 km, the path from Winnie the Pooh to Rabbit is 4 km and from Piglet to Winnie the Pooh is 3 km. How many kilometers will someone have to travel down the path from Eeyore to Rabbit?

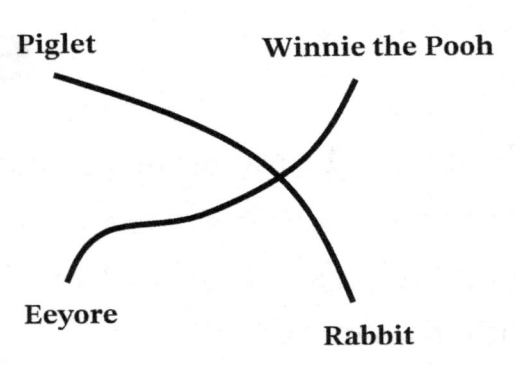

Problem 7. Petya goes up from the first floor to the fourth in 4 minutes. And Masha from the fourth floor to the seventh, in 3 minutes. Which of them will go the fastest from the first floor to the seventh and in how many minutes?

Problem 8. Twins Misha and Grisha lie at the same time only on Sunday. Other days, one lies and the other tells the truth. Misha said, "Today is Sunday". Grisha replied: "Sunday is tomorrow". What weekday is today?

ANSWER SHEET – EXAM 2

Problem 1	Answer
Problem 2	**Answer**
Problem 3	**Answer**
Problem 4	**Answer**

Mathematical Olympiads for Elementary School – *Problem Book*

Problem 5	Answer
Problem 6	Answer
Problem 7	Answer
Problem 8	Answer

Problems – Level 1

Exam 3

Problem 1. Mowgli, Bagheera the panther, Kaa the boa constrictor, Baloo the bear, and two wolf cubs met at the forest glade. How many legs are there in total?

Problem 2. The penguin is building an ice house. It remains to insert a part. Which?

Problem 3. Cut the pretzel in the picture into 6 pieces with two straight cuts.

Problem 4. Martians have three arms. The group from the Martian kindergarten lined up in pairs for a walk (see figure) According to the rules, each child must take each of her neighbors by the hand (left, right, forward or backward). How many hands free will the children in this group have after doing this?

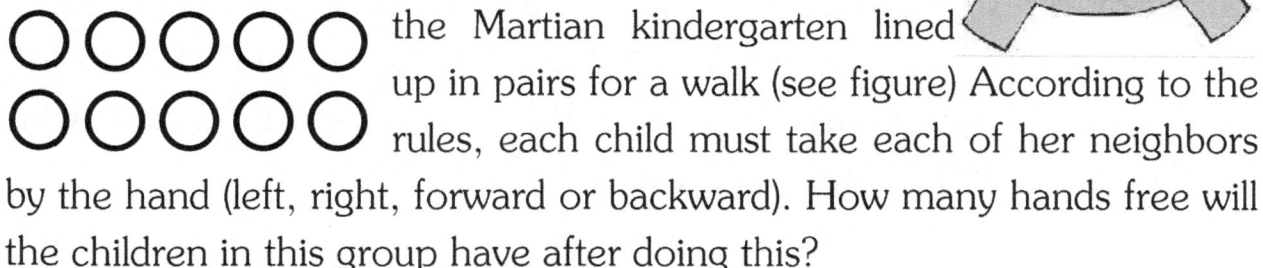

Problem 5. The ancient Romans, instead of the usual numbers 1, 2, 3, ... wrote the numbers in a different way: instead of 1 they wrote I, instead of 2 - II, instead of 3 - III, instead 4 - IV, instead of 5 - V, instead of 6 - VI, instead of 7 - VII. Vasya wrote the Roman numeral "5" on the sheet

of paper and Julia wrote the Roman numeral "4". They went with their sheets of paper to the mirror. Who has the highest number in the mirror: Vasya or Julia?

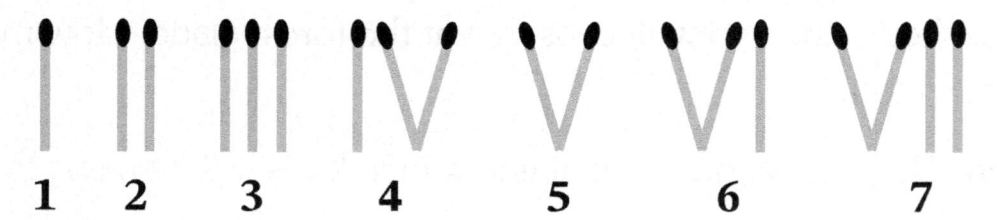

I	II	III	IV	V	VI	VII
1	2	3	4	5	6	7

Problem 6. Cheburashka runs every morning in a park with five oak trees. Draw the Cheburashka route and indicate in what order it passes the oaks, if it runs exactly once on each path. The direction of Cheburashka's movement is shown by arrows.

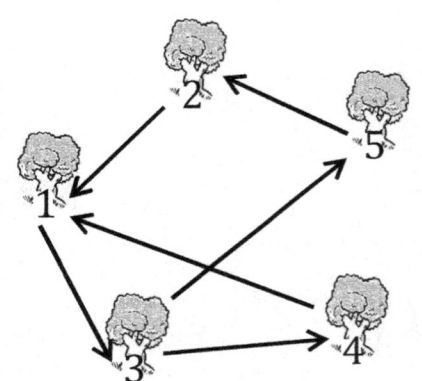

Problem 7. In a forest a group of mushrooms grow in a circle. In the forest glade, russulas always grow between two fly agarics. Vitalik counted without omitting anything, thus: "Two russulas, one fly agaric, seven russulas, one fly agaric". And Egor from another part of the circle: "Three russulas, a fly agaric, five russulas, a fly agaric". How many russulas are in the glade?

Problem 8. Krosh, Hedgehog, and Nyusha draw lots (three sticks of different lengths) for who would drive the car. Krosh said, "I have the shortest stick!" Nyusha said, "I'll drive!". Who drove the car if the one with the shortest stick did it and if both Krosh and Nyusha lied?

ANSWER SHEET – EXAM 3

Problem 1	Answer
Problem 2	**Answer**
Problem 3	**Answer**
Problem 4	**Answer**

Problem 5	Answer
Problem 6	**Answer**
Problem 7	**Answer**
Problem 8	**Answer**

Problems – Level 1

Exam 4

Problem 1. Petya has apples and Sasha has pears. If Petya trades one of her apples for two of Sasha's pears, then they will have the same number of fruits. How many more pears than apples are there?

Problem 2. Five gears are meshed with each other and rotate. If the upper left gear rotates clockwise as shown in the figure. In which direction does the lower left gear rotate?

Problem 3. On television, two episodes of a movie are broadcast on different channels on the same day. The first episode three times - at 12:00, 17:00 and 18:00 on the channel "AGA", and the second episode also three times - at 11:00, 15:00 and 17:00 on the channel "OGO". Egor wants to see the first episode first, and then the second. When should you turn on the television if each episode is exactly one hour long?

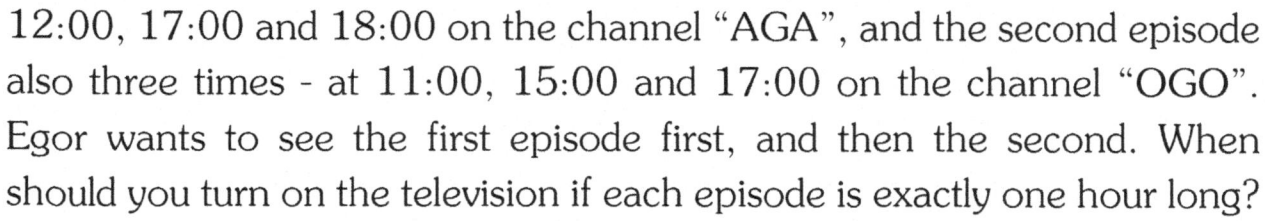

Problem 4. Gosha folded a square paper in half, then folded it in half again. Then he cut the resulting folded paper with two cuts, as in the figure. How many pieces of paper will Gosha find after unfolding it?

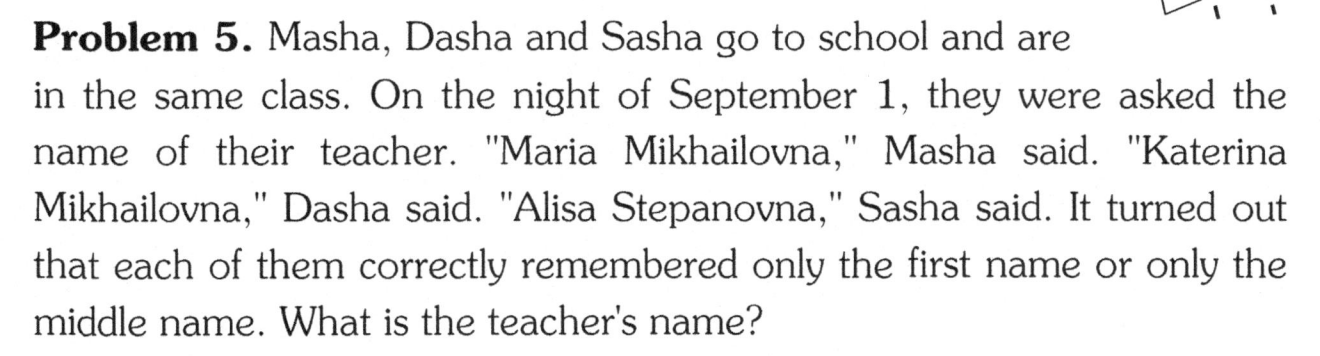

Problem 5. Masha, Dasha and Sasha go to school and are in the same class. On the night of September 1, they were asked the name of their teacher. "Maria Mikhailovna," Masha said. "Katerina Mikhailovna," Dasha said. "Alisa Stepanovna," Sasha said. It turned out that each of them correctly remembered only the first name or only the middle name. What is the teacher's name?

Problem 6. Passengers in 3 cars of a train carry 7 kittens, the third car has the least number of kittens and the second car has twice as many kittens as in the first car. How many kittens are in each car?

Problem 7. Losyash came up with building numbers with dominoes. For example, ⟦∙∙|∙⟧ is the number "21" and ⟦∙|∙∙⟧⟦∙∙∙|⟧ is 2350. Build the largest number possible with dominoes ⟦∙∙∙∙|∙⟧ y ⟦∙∙|∙∙⟧.

Problem 8. Mutta, Mokhovaya Beard and Polbotinka were eating ice cream. "Polbotinka ate more than everyone!" - Mutta said. "No, I ate less than Mokhovaya Beard", Polbotinka objected. "Polbotinka and I ate alike", Mokhovaya Beard said in a conciliatory tone. Who ate less if it is known that everyone lied?

ANSWER SHEET – EXAM 4

Problem 1	Answer
Problem 2	**Answer**
Problem 3	**Answer**
Problem 4	**Answer**

Problem 5	Answer
Problem 6	Answer
Problem 7	Answer
Problem 8	Answer

Problems – Level 1

Exam 5

Problem 1. Now Tanya, Mana and Anya are 12 years old in total. How old will they be in 2 years?

Problem 2. In the number 798, all the numbers are different. What number closest to this has the same property?

Problem 3. Petya brought her dachshund dog Dina to the dog show. If you photographed the dachshunds in the order the spots were assigned. Where did Dina go? (Petya was photographing against the sun, so only silhouettes were obtained.)

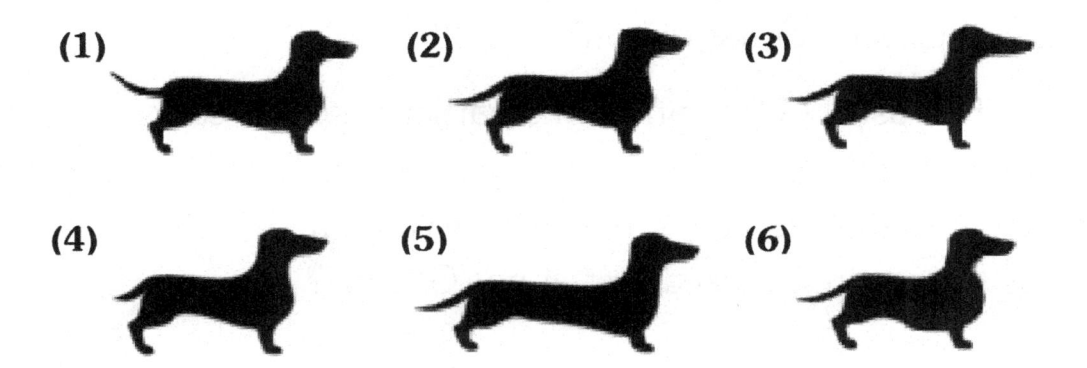

Dina

(1) (2) (3)

(4) (5) (6)

Problem 4. Draw 10 shapes in a row: circles and squares so that next to each circle there are only squares, and next to each square there is a circle and a square.

Problem 5. Yegor sketched a mountain with a layer of snow with the help of matches. rearrange 2 matches so that exactly 3 triangles are visible in the figure (there should be no additional matches).

83

Problem 6. Vanya received a 9-cell chocolate bar as shown in the figure. If you want to eat two cells of the chocolate bar so that the rest does not fall apart. How many ways could she do it?

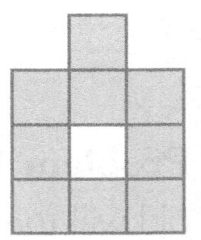

Problem 7. A beetle moves along a 1 meter long ribbon (as in the picture). It begins to move from the starting point 2 *cm* from one edge of the ribbon and moves strictly in the middle of the ribbon, without going sideways or going back. How much distance will the beetle travel when it moves toward the arrival point 3 *cm* from the other edge of the ribbon?

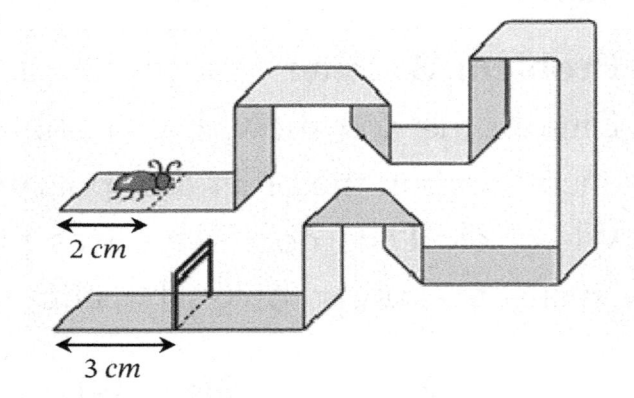

Problem 8. Vasya, Grisha and Dima participated in a car race with three cars: blue, red and yellow. Dima's car reached the finish line just after the yellow car and the red car, just after Vasya's car. Whose car and what color was the car that arrived first, if not Grisha's?

Problems – Level 1

ANSWER SHEET – EXAM 5

Problem 1	Answer
Problem 2	**Answer**
Problem 3	**Answer**
Problem 4	**Answer**

Problem 5	Answer
Problem 6	Answer
Problem 7	Answer
Problem 8	Answer

Problems – Level 1

Exam 6

Problem 1. A mouse made a big hole in a sweater. Baba-Manya wove various patches. Which is the most suitable for mending the sweater?

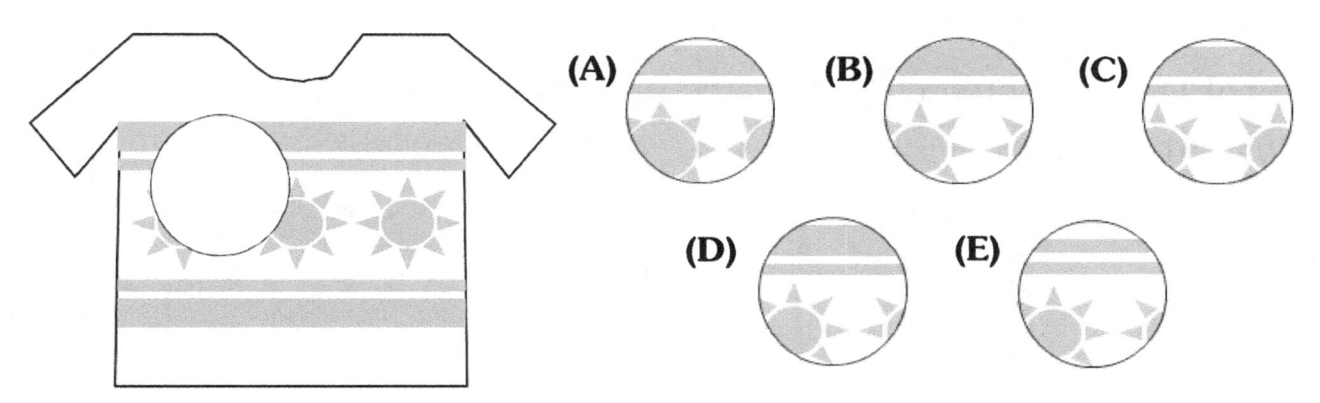

Problem 2. I have three friends: Anton, Borya and Kolya. Yesterday I played with Borya and Anton. One of them is 8 years old and the other 9 years old. And today I walked with Anton and Kolya. One of them is 10 years old and the other 8 years old. How old is each of my friends?

Problem 3. Instead of the asterisks, put some numbers to get the correct equality

$$2 * - * - * = 3$$

Problem 4. A beetle is in the cell of a board (as in the figure). If he walked through 2 cells and stopped at the third. Indicate which cell it could be in.

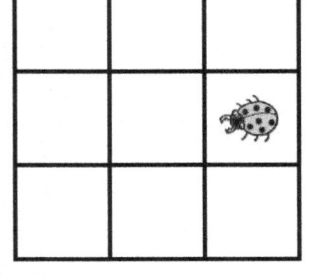

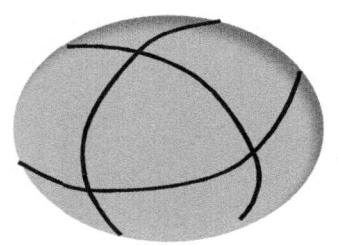

Problem 5. A watermelon was cut with three straight cuts as shown in the figure. How many pieces of peel were obtained after eating the watermelon?

Problem 6. Vasya, Gosha, and Kazimir painted a white triangle, a gray circle, and a black square with oil paints (each child drew a shape). It is known that Vasya painted later than Kasimir, and Vasya's figure is not white. Who drew what shape?

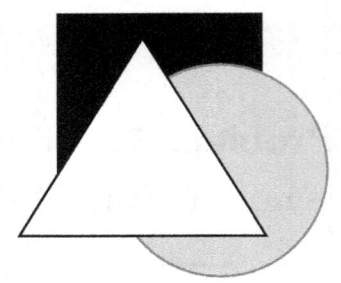

Problem 7. Some children came up with the idea of adding numbers with matches:

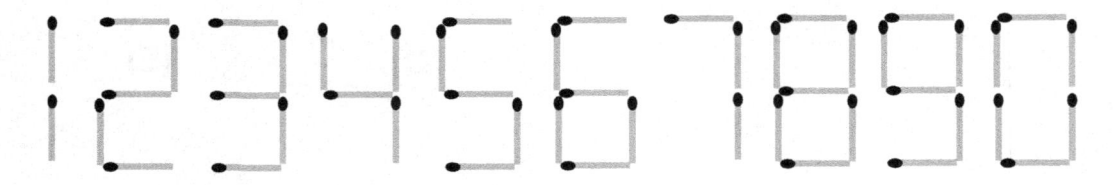

1234567890

Anya posted the following wrong equality. Move 2 matches so the equality is true:

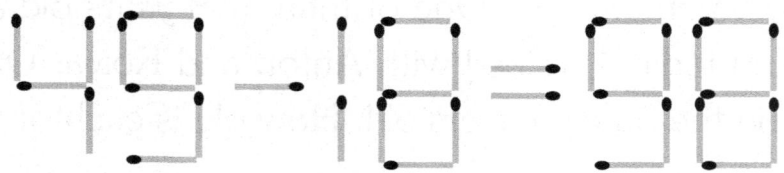

49 - 18 = 98

Problem 8. Three friends Tikhon, Yegor and Vitalik exchanged toys. Vitalik started playing with a fire truck. The owner of the dump truck liked the excavator. The owner of the fire truck took the dump truck. Determine which toy is whose, if the fire truck is known not to be from Tikhon.

Problems – Level 1

ANSWER SHEET – EXAM 6

Problem 1	Answer
Problem 2	Answer
Problem 3	Answer
Problem 4	Answer

Problem 5	Answer
Problem 6	Answer
Problem 7	Answer
Problem 8	Answer

Problems – Level 1

Exam 7

Problem 1. Petya drew several maple leaves and took one home. Which of the following pictures did Petya make?

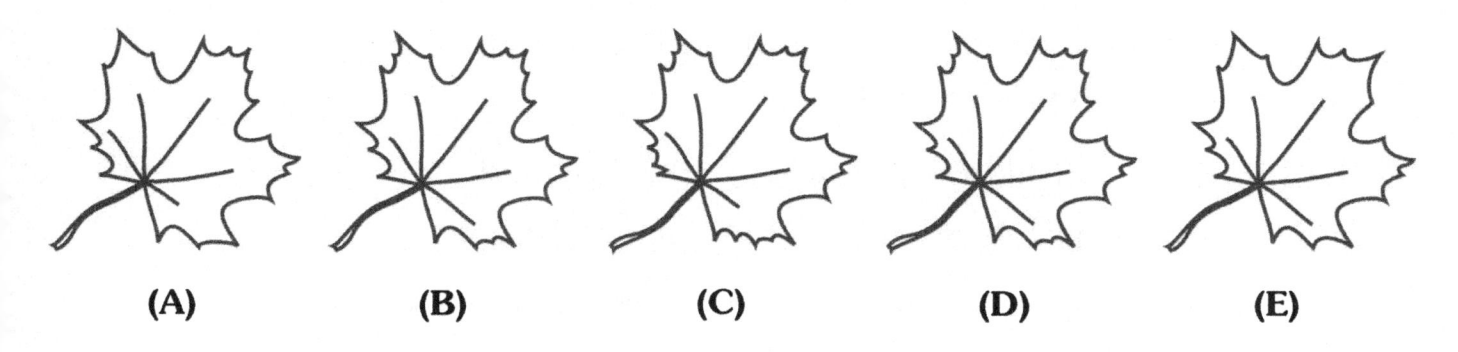

(A)   (B)   (C)   (D)   (E)

Problem 2. Vasya spends more time eating breakfast than Petya brushing her teeth and washing her ears. And the Murka spends the same time washing as Vasya eating breakfast. Who will finish washing faster, Murka or Petya?

Problem 3. Replace the letters with numbers from 1 to 7, so that the inequality is true. If different letters represent different numbers.

$$S < N < E > W < I < N > K > A$$

Problem 4. Masha, Katya, Tikhon, Yegor and Sveta line up for ice cream. It is known that Yegor is in front of Tikhon and Katya is behind Masha and in front of Sveta. What is the order of the children in line if there are no two girls together?

Problem 5. Krosh has two clocks. One of them is 2 hours ahead, and the other is 1 hour behind. What time is it now, if the clocks show the time shown in the figure to the side?

Problem 6. Petya and Vasya played the naval battle on a 6 × 6 board. Vasya painted over the cells where Petya no longer has ships. Petya still has a three-deck ship ⊏⊐. Which cell should Vasya attack to safely hit Petya's ship?

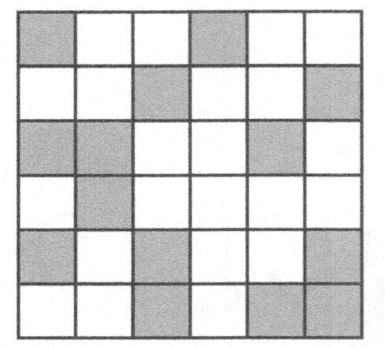

Problem 7. Tikhon represents the natural numbers of a single digit with the help of matches:

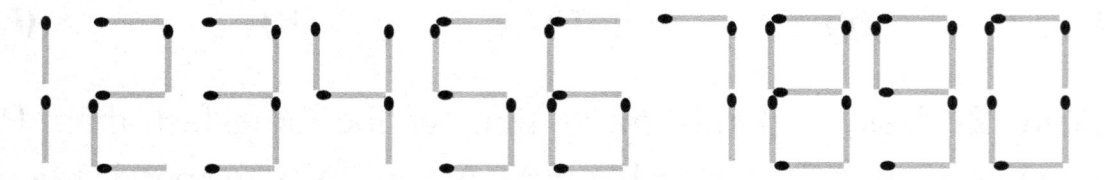

He also established the correct equality 2 + 6 = 9 − 1. Move 2 matches to get another correct equality.

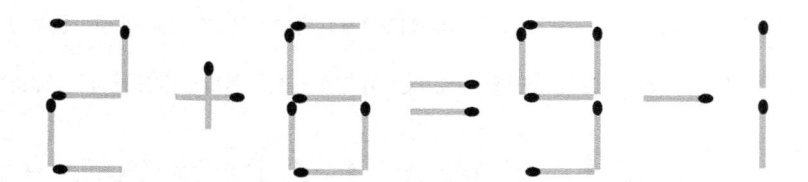

Problem 8. Vika, Nastya and Sonya are learning to count. Each of them has 1 or 2 candies. Sonya said, "We have at least 5 candies". Nastya: "Vicki and I have equal amounts". Vika: "I have more candy than Sonya". It turned out that they were all wrong. How many candies does each have?

ANSWER SHEET – EXAM 7

Problem 1	Answer
Problem 2	Answer
Problem 3	Answer
Problem 4	Answer

Problem 5	Answer
Problem 6	Answer
Problem 7	Answer
Problem 8	Answer

Problems – Level 1

Exam 8

Problem 1. Color the four circles in three different colors so that two adjacent circles are different colors.

Problem 2. Indicate which of the 4 cube arrangements shown here are

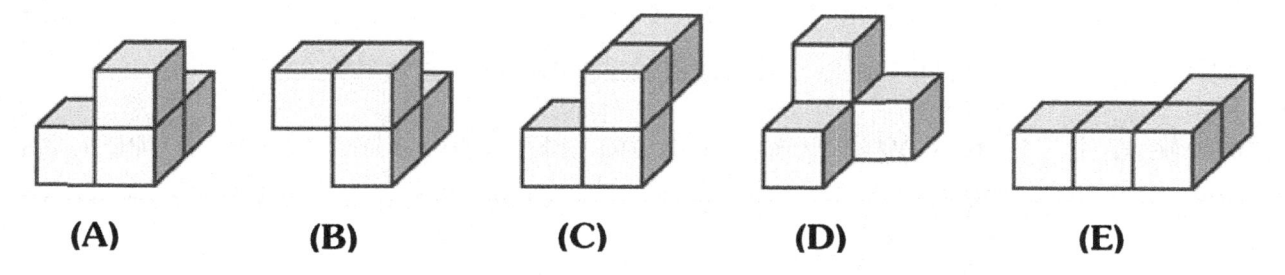

(A)   (B)   (C)   (D)   (E)

Problem 3. Varina's father is named Nikita Andreevich and her grandfather is Eduard Vasilyevich. What is the middle name of Varina's mother?

Problem 4. Jung is training to tie knots. The image shows his five attempts. What knots will be tied if the rope is pulled at the ends?

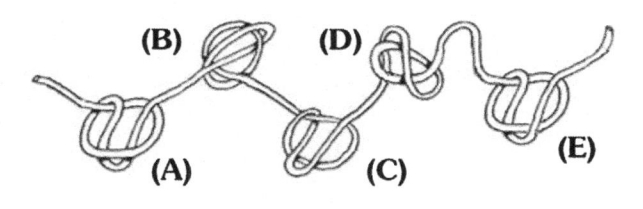

Problem 5. Masha forms numbers with matches. She places the number "5" in front of a mirror in every possible way and looks at it from different sides. What one- and two-digit numbers can be viewed this way?

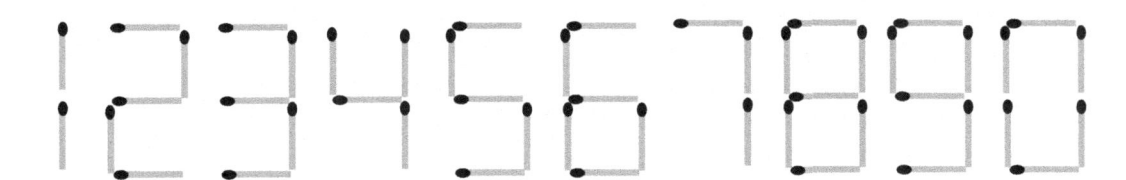

Problem 6. To play "Twister" a field of 12 cells with 4 different colors is used: red (R), yellow (Y), green (G) and blue (B) (see figure). Murka stretches across the field so that only 4 cells of different colors remain free, which are not adjacent. Indicate which cells Murka occupies.

R	G	B	Y
B	Y	R	G
R	G	B	Y

Problem 7. Piglet planted 10 acorns. All but three grew oak trees. All but two oaks have acorns. In all but one oak with acorns, the acorns have no flavor. How many oaks with tasteless acorns are there?

Problem 8. One day it was snowing all night. That night, three cars arrived at different times and parked near a house (as shown in the figure). Determine in what order the cars arrived.

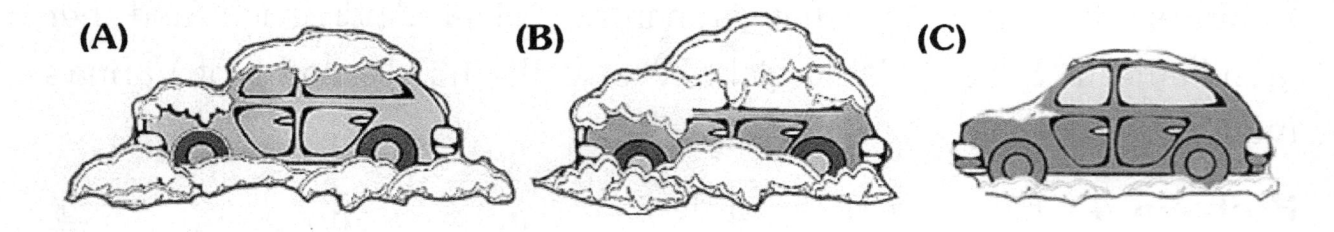

(A) (B) (C)

Problems – Level 1

ANSWER SHEET – EXAM 8

Problem 1	Answer
Problem 2	**Answer**
Problem 3	**Answer**
Problem 4	**Answer**

Mathematical Olympiads for Elementary School – *Problem Book*

Problem 5	Answer
Problem 6	Answer
Problem 7	Answer
Problem 8	Answer

Problems – Level 1

Exam 9

Problem 1. Replace the letters with digits (if different letters represent different digits) to get the correct equalities:

$$P - O = B - E = D - I = T + E = L + I$$

Problem 2. Three parrots have 9 nuts together. Red is 1 more than green and blue is 1 less than green. How many nuts does each of the parrots have?

Problem 3. Winnie the Pooh flies with 7 balloons to look for honey and now he can't go down. Piglet has a pistol, from which he can only shoot 2 times. If he wants to pop several balloons in one shot. Show how Piglet can help Winnie the Pooh get him down by shooting all the balloons.

Problem 4. Alena wants to untangle her ribbons. How many are there?

Problem 5. Buratino has four of the same coins. If he put them in 4 bags (see figure below) so that each bag contains a different number of coins. Indicate how the coins are distributed in the bags.

Problem 6. Natasha formed digits with matches and posted an incorrect equality as shown below.

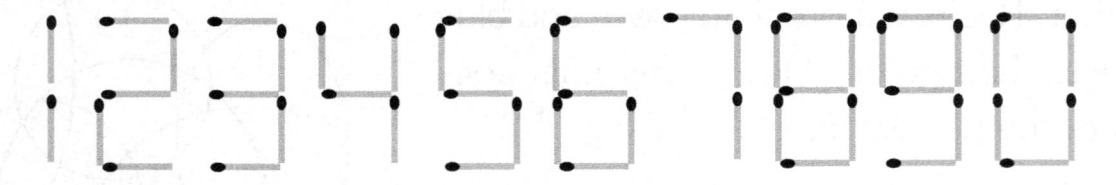

$$1234567890$$

Move 2 matches to get the correct equality:

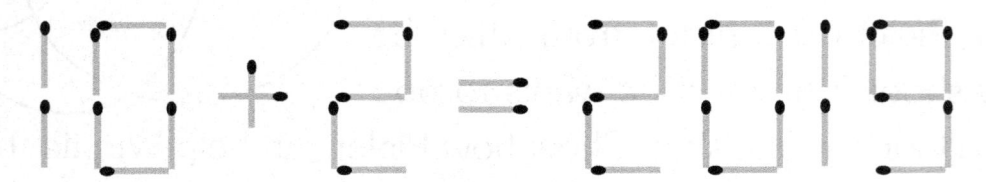

$$10 + 2 = 2019$$

Problem 7. Go through the maze by turning left three times and right three times (in any order).

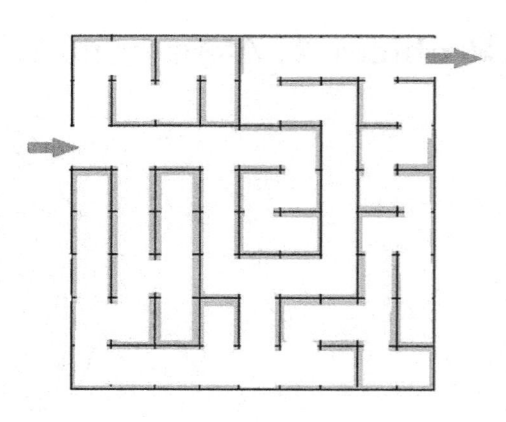

Problem 8. Five soldiers are standing in a row: Anton, Nikolay, Fedor, Alexey, Ivan. The colonel ordered two neighboring soldiers to break ranks, and then did the same with those whose names begin with the same letter. Finally, a soldier remained in line. What is his name?

ANSWER SHEET – EXAM 9

Problem 1	Answer

Problem 2	Answer

Problem 3	Answer

Problem 4	Answer

Problem 5	Answer
Problem 6	**Answer**
Problem 7	**Answer**
Problem 8	**Answer**

Problems – Level 1

Exam 10

Problem 1. There are six brothers in the Zephyr family, two of them are twins, very similar to each other, like two drops of water. What are the names of these twins?

Max   *Wax*   *Pax*   *Rex*   *Fex*   *Kex*

Problem 2. Masha formed a word with letters contained in gray and white cards. Grisha noted that it is possible to exchange white and gray cards so that the colors of the cards alternate. What cards must be exchanged?

O L I M P I A D A

Problem 3. On Christmas Eve, Misha made Christmas trees from a honeycomb (as on the left). From the figure on the right, she cut two Christmas trees. How did he do it?

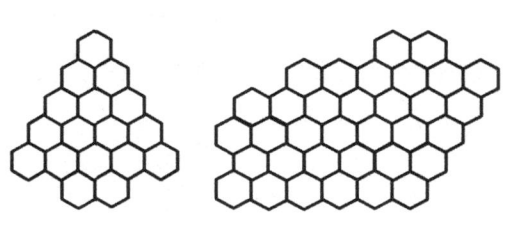

Problem 4. There are several 🌲 trees and ● light poles in the park. The light pole illuminates the trees strictly horizontally (↔) and vertically (↕). The tree closest to a light pole blocks the rest of the light in that direction. Mark which tree or trees A) are not illuminated with "O"; B) illuminated by exactly two light poles with "×"

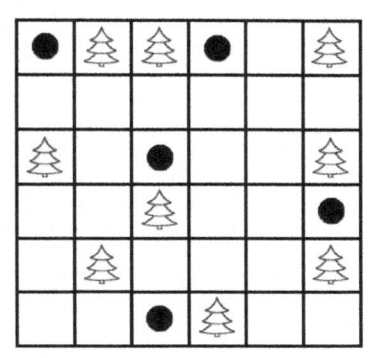

Problem 5. Misha took three dominoes [1|2] [1|6] [2|5] and put them in a scheme as shown in the figure. It turned out that the sum of points in two vertical lines and one horizontal is the same. Show how Misha arranged the dominoes.

Problem 6. The magic white sheet of paper changes its color to black in the places where the white parts touch. The square sheet was folded twice (as in the figure), pressed and cut diagonally. How many pieces that are completely black on both sides are there?

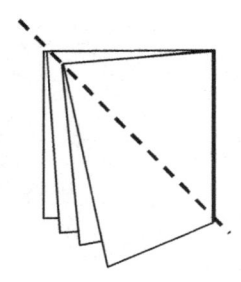

Problem 7. There are four houses along a straight street: blue, yellow, green, and red (in that order). The fox does not live in the red house. And the neighbors of the hare are a bear and a hedgehog. Who lives and where if the fox is not next to the hedgehog?

Problem 8. Snow was falling at night. In the morning, Fyodor in sneakers, Sharik in boots, and Matroskin in felt boots walked through the fresh snow. In what order were they?

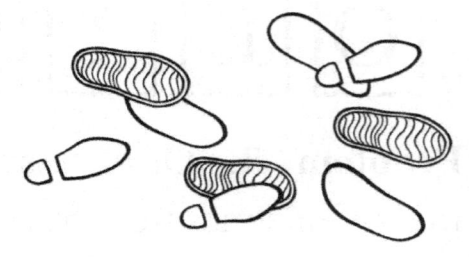

Problems – Level 1

ANSWER SHEET – EXAM 10

Problem 1	Answer
Problem 2	**Answer**
Problem 3	**Answer**
Problem 4	**Answer**

Problem 5	Answer
Problem 6	Answer
Problem 7	Answer
Problem 8	Answer

Answers: Level 1

Answers – Level 1

Exam 1

1. Nyusha.
2. Option (C).
3. 16 holes.
4. 2 green cells.
5. See the next figure:

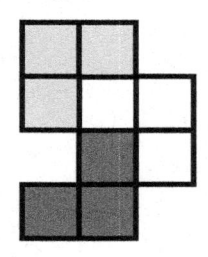

6. He should go into the thicket.
7. Winnie the Pooh.
8. Andrey loves oranges and Vera loves apples.

Exam 2

1. Pavel Ivanovich.
2. The solution is the following:

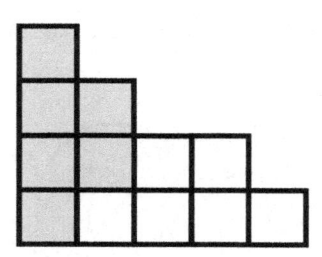

3. the picture of apple number 7 is missing. The figure shows the correspondences for the remaining fruits.

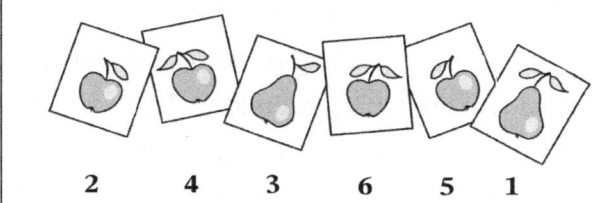

2 4 3 6 5 1

4. See the table below:

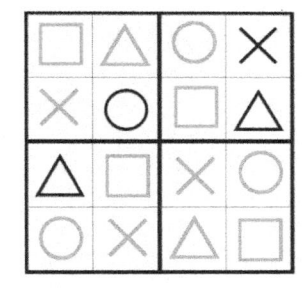

5. The Honda is blue, the Mercedes is yellow, and the Audi is red.
6. 8 kilometers.
7. Masha is 2 minutes faster.
8. Today is Saturday.

Answers – Level 1

Exam 3

1. A total of 18 legs.
2. Option (D).
3. The figure shows one of the possible cut options.

4. There will be four hands free.
5. Julia has a higher number in the mirror.
6. The Cheburashka route is: 3-4-1-3-5-2-1.
7. There is 12 russulas in the glade.
8. Hedgehog drove the car.

Exam 4

1. There are 2 more pears than apples.
2. Rotate clockwise.

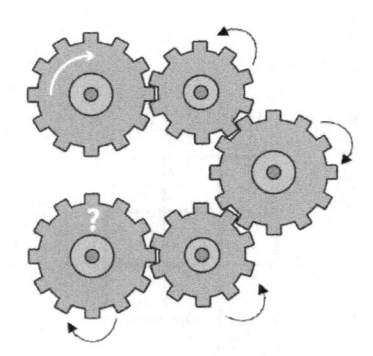

3. Tune in "AGA" at 12 o'clock and then tune in "OGO" at 15 or 17 o'clock.
4. Gosha will find 5 pieces after unfolding such a paper.
5. The teacher's name is Alisa Mikhailovna.
6. In the first car there are 2 kittens, in the second – 4 kittens, in the third – 1 kitten.
7. This number is 6243.
8. Mokhovaya Beard ate less than the others.

Answers – Level 1

Exam 5

1. 18 years old.
2. 796.
3. The second image.

4. The solution is as follows:

○□□○□□○□□○

5. By moving the two matches at the top and placing them at the bottom, we can then see 2 large and 1 small triangles.

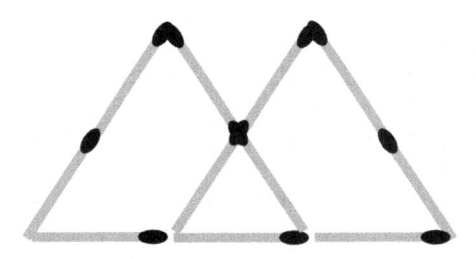

6. 7 ways.
7. 95 cm.
8. Vasya's yellow car arrived first.

Exam 6

1. Option (D).
2. Anton is 8 years old, Borya is 9 years old, Kolya is 10 years old.
3. $21 - 9 - 9 = 3$ or $20 - 9 - 8 = 3$ or $20 - 8 - 9 = 3$.
4. In the figure, the possible cells are shaded gray.

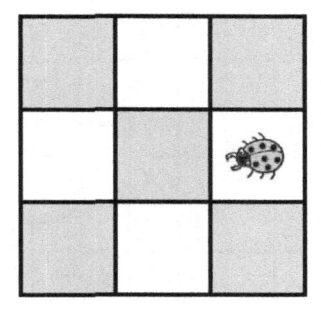

5. 8 pieces.
6. Kasimir drew the black square, Vasya drew the gray circle, Gosha drew the white triangle.
7. The correct equality is 40+18=58.

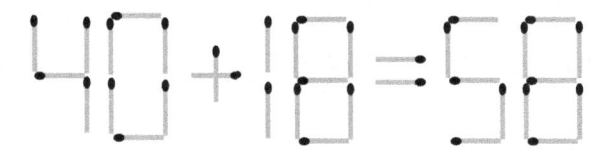

8. Fire truck - Yegor, dump truck - Tikhon, excavator - Vitalik.

Answers – Level 1

Exam 7

1. Option (D).
2. Petya is faster.
3. For example, 1 < 6 < 7 > 2 < 3 < 6 > 5 > 4.
4. Masha, Egor, Katya, Tikhon, Sveta.
5. It's 6 o'clock now.
6. The required cell is marked in the figure.

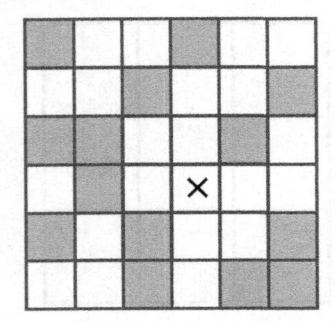

7. The correct equality is 2+5=6+1.

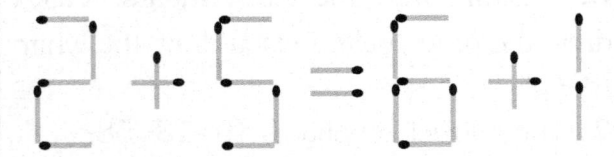

8. Sonya has 1 candy, Nastya has 2 candy, and Viki - has 1 candy.

Exam 8

1. For example, red - blue - yellow - red.
2. The options (A) and (D) are identical.
3. The middle name is Eduardovna. Since Nikita Andreevich's middle name is not Eduardovich, Eduard Vasilyevich is not his father. But he is Varina's grandfather. This means that this grandfather is the father of Varina's mother, from which we get her middle name.
4. Knots (B) and (E) will be tied.
5. Masha will be able to see the numbers 5, 2, 3, 25, 52.
The dotted line shows the position of the mirror:

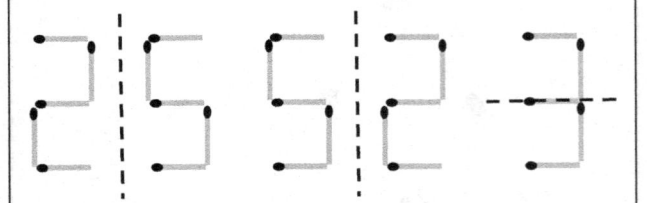

6. Two symmetrical variants are possible:

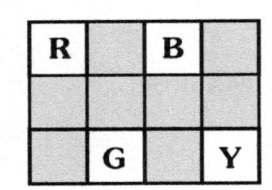

R		B
	G	Y

G		Y
R	B	

7. There are 4 oaks with tasteless acorns.
8. Car B came first, Car A came second, Car C came third.

Answers - Level 1

Exam 9

1. For example, 9 – 3 = 8 – 2 = 7 – 1 = 4 + 2 = 5 + 1.

2. Green has 3 nuts, blue has 2 nuts, red has 4 nuts.

3. Two shots in the shaded areas will suffice, one in each.

4. 3 ribbons. In the figure, each ribbon is highlighted in its own color.

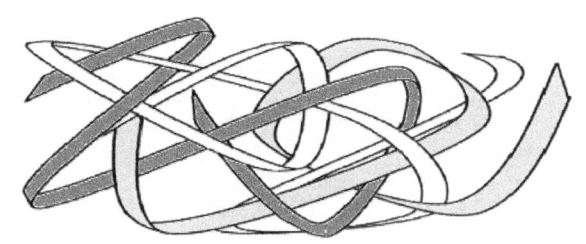

5. An example is shown in the next figure:

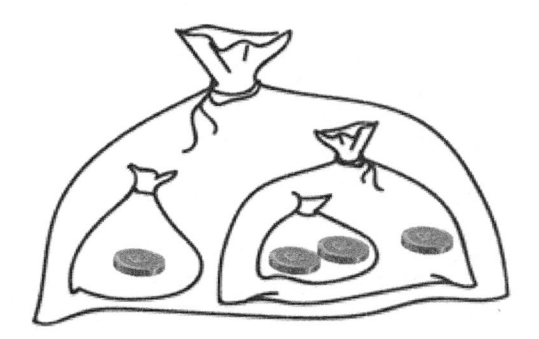

6. An option is shown in the next figure:

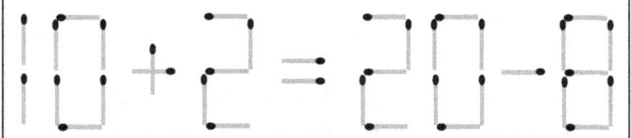

7. An option is shown in the next figure:

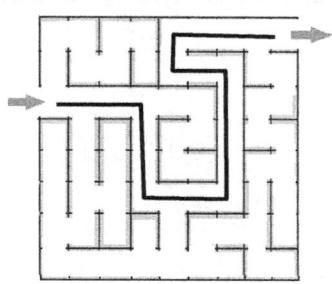

8. The name of the remaining soldier is Ivan.

Answers - Level 1

Exam 10

1. Wax and Kex.

2. "P" and "I" must be interchanged

O L I M P I A D A

3. An example is as shown as follows:

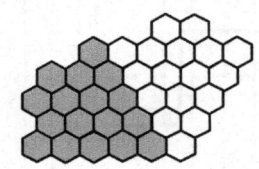

4. The trees are marked on the image.

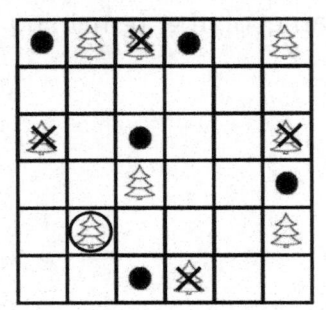

5. Un example is shown in the figure:

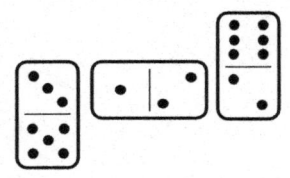

6. 2 pieces.

7. The fox is in the blue house, the bear is in the yellow one, the hare is in the green one, and the hedgehog is in the red one.

8. First Matroskin, then Fedor, and last Sharik.

Problems: Level 2

Problems – Level 2

Exam 1

Problem 1. There were cherries on a plate. Misha ate half of all the cherries and 17 more cherries. Likewise, Misha tossed the remaining spoiled cherry. How many cherries did Misha eat?

Problem 2. A triangular portion of a large gift-wrapping paper has been cut out. Petya found 6 different pieces. Which of the pieces belongs to this paper?

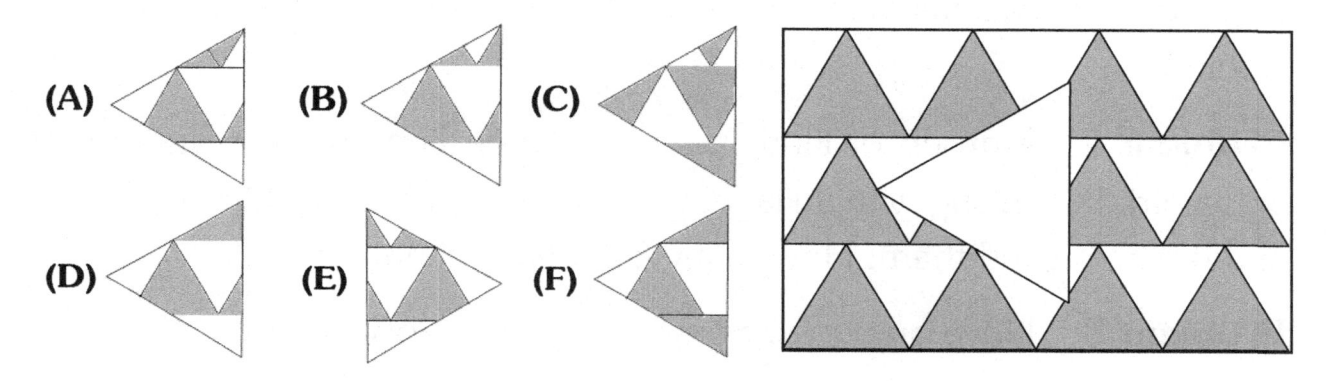

(A) (B) (C) (D) (E) (F)

Problem 3. The wolf Vasya, the tiger Tosha, the hare Styopa and the zebra Zoya came to the carnival wearing wolf, tiger, hare and zebra masks. Also, Styopa and Zoya wear predator masks, while the owners of the zebra and tiger masks do not have stripes. Who wears what mask?

Problem 4. In a 9-cell square, Olya painted more than 7 cells with a yellow pencil, and then Ira painted more than 5 cells with a blue pencil. It turned out that all the cells in the square are painted. If you paint the cell with yellow and blue pencils, you get green. How many green cells are in the square?

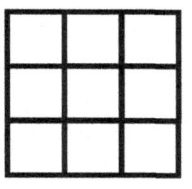

Mathematical Olympiads for Elementary School – *Problem Book*

Problem 5. Two dice were placed one on top of the other and photographed (see figure below). A) How many points in total are there on the photographed faces? B) How many points in total are there on all the other faces?

Problem 6. A scale was taken to the Enchanted Forest and everyone ran to be weighed. It turned out that Winnie the Pooh and Christopher Robin together weigh more than Piglet and the donkey Eeyore together. And Eeyore weighs more than Winnie the Pooh and Little Roo put together. Who weighs more, Piglet and Little Roo together or Christopher Robin?

Problem 7. A group of kindergarten children lined up in pairs, a boy and a girl. Ilya, along with Julia, counted 5 boys in front of him and Julia, and 4 girls behind them. How many children are in the group?

Problem 8. "I love oranges", Sonya said. "No, I love oranges. You love apples", Vera said. "I don't like oranges", Andrei said. It is known that there is at least one child who likes each fruit and that neither child likes both fruits at the same time. Which child loves apples and which child loves oranges, if they all lie?

ANSWER SHEET – EXAM 1

Problem 1	Answer
Problem 2	**Answer**
Problem 3	**Answer**
Problem 4	**Answer**

Problem 5	Answer
Problem 6	**Answer**
Problem 7	**Answer**
Problem 8	**Answer**

Problems – Level 2

Exam 2

Problem 1. Between some digits, put an equal sign and an arithmetic sign to get the correct equality:

1 2 3 4 2 2

Problem 2. Cut the checkered figure shown below into three equal parts.

Problem 3. Andryusha photographed reflections of fruits in the mirror. And then he lost a picture. Which picture is Andryusha missing now?

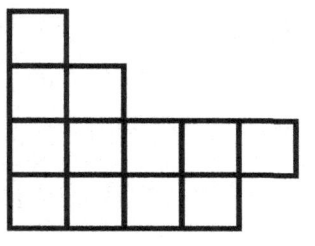

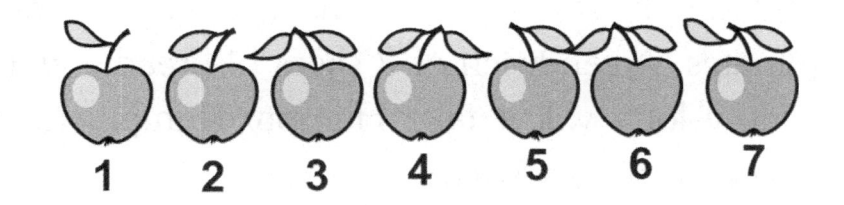

1 2 3 4 5 6 7

Problem 4. In the table below, arrange the circles, triangles, squares, and crosses so that in each column and each row, as well as each highlighted small square, there are all four figures.

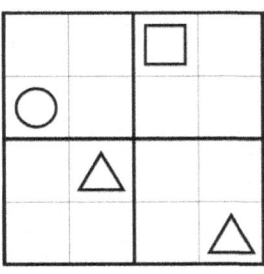

Problem 5. Three cars participating in a race came out in this order: yellow, red, blue. They reached the finish line in this order:

"Honda", "Mercedes", "Audi". At the same time, no car finished the same way it started. What color is each car brand, if Audi is not yellow?

Problem 6. Paths have been built in the Enchanted Forest (see image). It turned out that the path from Piglet to Eeyore Donkey is 7 km, the path from Winnie the Pooh to Rabbit is 4 km and from Piglet to Winnie the Pooh is 3 km. How many kilometers will someone have to travel down the path from Eeyore to Rabbit?

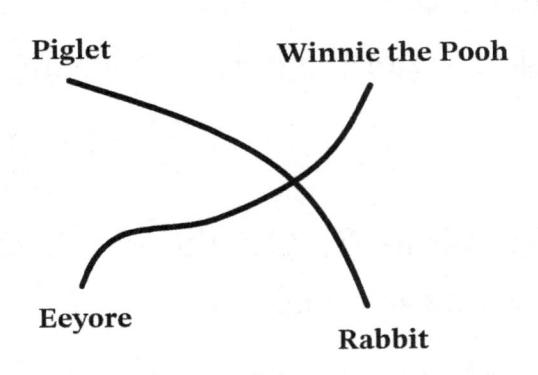

Problem 7. In any Humpty Dumpty sandwich, the sausage and bread slices alternate. Humpty eats a sandwich of a slice of bread and 2 of sausage in 4 minutes. A sandwich of 2 slices of bread and 1 of sausage, in 5 minutes. How long will it take Humpty Dumpty to eat a 5-slice sausage and 4-slice bread sandwich?

Problem 8. Physicists and chemists attended a scientific conference. They are all divided into theoretical and experimental. It is known that theorists always lie and experimentalists always tell the truth. The next speaker began his speech with the statement "I am a theoretical chemist". What is really the speaker?

ANSWER SHEET – EXAM 2

Problem 1	Answer

Problem 2	Answer

Problem 3	Answer

Problem 4	Answer

Problem 5	Answer

Problem 6	Answer

Problem 7	Answer

Problem 8	Answer

Problems – Level 2

Exam 3

Problem 1. Nikita and her sister Olya live in their apartment together with mom, dad, dog, three cats, and five goldfish. How many legs are there in total?

Problem 2. Replace letters in the expression M + A > T < E < M > A + T > I > K + A letters with numbers from 1 to 6 so that you get the correct inequalities (If different letters represent different numbers).

Problem 3. Cut the pretzel in the picture into 7 pieces with two straight cuts.

Problem 4. Martians have five arms. The group from the Martian kindergarten lined up in pairs for a walk (see figure) According to the rules, each child must take each of her neighbors by the hand (left, right, forward or backward). How many hands free will the children in this group have after doing this?

Problem 5. The ancient Romans, instead of the usual numbers 1, 2, 3, ... wrote the numbers in a different way: instead of 1 they wrote I, instead of 2 - II, instead of 3 - III, instead 4 - IV, instead of 5 - V, instead of 6 - VI, instead of 7 - VII. Vasya wrote the operation "VI - V = I" on the sheet of paper and Julia wrote the operation "II = VI - IV". They went with their sheets of paper to the mirror. Who will get the correct equality in the mirror: Vasya or Julia?

Problem 6. Cheburashka runs every morning in a park with six oak trees. Draw the Cheburashka route and indicate in what order it passes the oaks, if it runs exactly once on each path. The direction of Cheburashka's movement is shown by arrows.

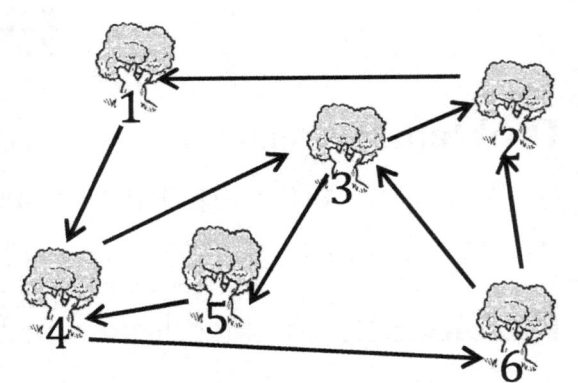

Problem 7. In a forest a group of mushrooms grow in a circle. In the forest glade, russulas always grow between two fly agarics. Vitalik counted without omitting anything, thus: "Two russulas, one fly agaric, seven russulas, one fly agaric". And Egor from another part of the circle: "Three russulas, a fly agaric, five russulas, a fly agaric". How many russulas are in the glade?

Problem 8. Krosh, Hedgehog, and Nyusha draw lots (three sticks of different lengths). Krosh said, "I have a normal stick!", Hedgehog said, "I have it shorter than Krosh's", and Nyusha said, "And I have it shorter than Hedgehog's!". Who has the longest stick if everyone is lying?

Problems – Level 2

ANSWER SHEET – EXAM 3

Problem 1	Answer

Problem 2	Answer

Problem 3	Answer

Problem 4	Answer

Problem 5	Answer
Problem 6	Answer
Problem 7	Answer
Problem 8	Answer

Problems – Level 2

Exam 4

Problem 1. In the expression A + P + E + L + M + C + I + H + N = EE, replace the same letters with the same digits and different letters with different digits to get the correct equality.

Problem 2. Six gears are meshed with each other and rotate. If the upper left gear rotates clockwise as shown in the figure. In which direction does the lower left gear rotate?

Problem 3. On television, two episodes of a movie are broadcast on different channels on the same day. The first episode three times - at 12:30, 14:00 and 16:00 on the channel "AGA", and the second episode also three times - at 10:30, 13:00 and 14:30 on the channel "OGO". Egor wants to see the first episode first, and then the second. When should you turn on the television if each episode is exactly 90 minutes long?

Problem 4. Cut the shape on the right into three equal parts.

Problem 5. Arthur, Misha, Kolya, and Vasya signed up for a judo class. In the evening they were asked what the name of their instructor was. "Semyon Yegorovich Zadornov", Arthur said. "Semyon Pavlovich Veselovsky", Misha said. "Alexander Pavlovich Smekhov", Kolya said. "Efim Petrovich Zadornov", Vasya said. It turned out that each of them correctly remembered only the first name, or only the middle name, or only the last name. What is the name of the instructor?

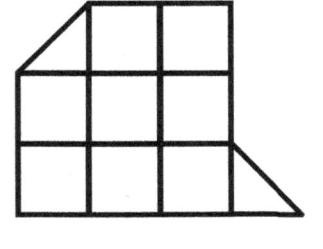

Problem 6. Losyash came up with building numbers with dominoes. For example, [domino] is the number "21" and [dominoes] is 2350. Build the largest number possible with dominoes [domino], [domino] y [domino].

Problem 7. A red rose, a white lily, and a red carnation grow in a row in red, white, and yellow pots on the windowsill. It is known that, the red flowers are next to each other. The yellow pot is not at the ends. The rose is next to the white pot. And no flower grows in a pot of the same color. Determine how the flowers are arranged on the windowsill.

Problem 8. The children in a class are going for a walk. If they get together in pairs, a boy and a girl, then three girls will not have a partner. And if two girls get together with each boy, in the end there will be two boys left over. How many boys and how many girls will there be in this class?

ANSWER SHEET – EXAM 4

Problem 1	Answer
	|//
Problem 2	Answer
Problem 3	Answer
Problem 4	Answer

Mathematical Olympiads for Elementary School – Problem Book

Problem 5	Answer

Problem 6	Answer

Problem 7	Answer

Problem 8	Answer

Problems – Level 2

Exam 5

Problem 1. Sasha, Misha, Nina, Vasya and another child from kindergarten cut out the letters of their names from sheets of paper. It turned out that they cut 7 letters A, 4 letters S, 3 letters I and N each, 2 letters V and H each, and 1 letter M and Y each. What is the name of the fourth child?

Problem 2. A piece of rope was bent in half and again in half. And then the resulting skein is cut in the middle. How many pieces of rope did you get?

Problem 3. In the number 16798, all the numbers are different. What number closest to this has the same property?

Problem 4. A figure of matches is presented as in the image below. You can see 4 squares: 3 small and 1 large. Move 2 matches so that only 3 squares can be seen. (There should be no extra matches)

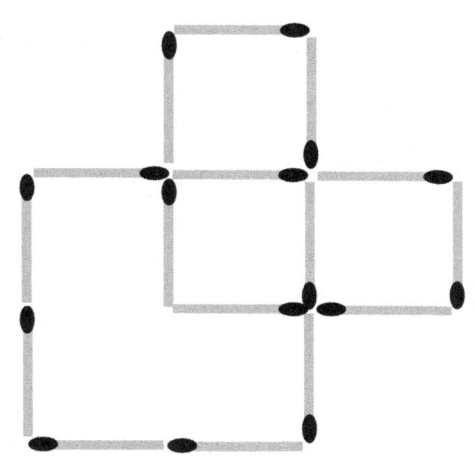

Problem 5. Kopatych went to bed and slept for 14 hours. Indicate what time the clock showed, when he went to bed, and what time when he woke up.

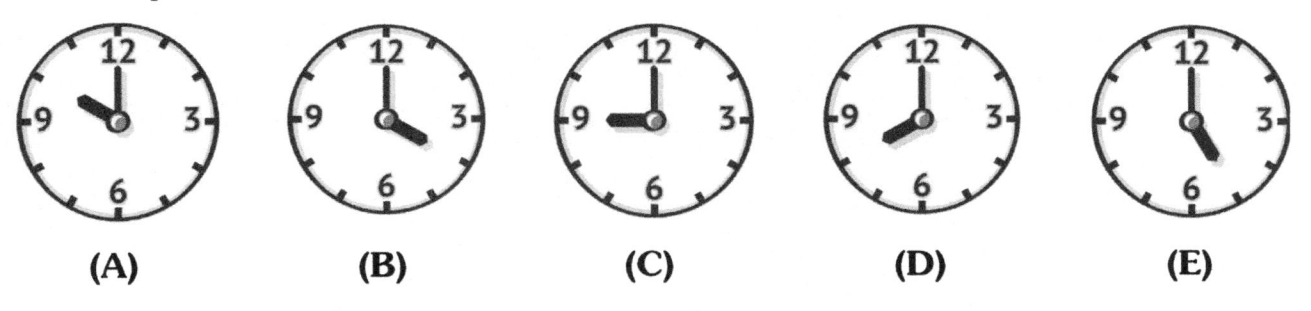

(A) (B) (C) (D) (E)

Problem 6. Martian cats are similar to terrestrial cats, they also have 4 paws, but the number of toes on each paw can be different. Aelita's cat has a total of 8 toes on the left paws, 9 on the right paws, and 11 on the front paws. How many toes does this cat have on its hind paws in total?

Problem 7. Four dice are placed as shown in the figure. It turned out that the sum of the points of the visible faces (top and side) is 34. What is the sum of the points of the non-visible faces? (A die is a cube with points 1 through 6 drawn on the faces)

Problem 8. Vasya, Grisha and Dima participated in a car race with three cars: blue, red and yellow. Dima's car reached the finish line just after the yellow car and the red car, just after Vasya's car. Whose car and what color was the car that arrived first, if not Vasya's and Grisha doesn't have a blue car?

ANSWER SHEET – EXAM 5

Problem 1	Answer

Problem 2	Answer

Problem 3	Answer

Problem 4	Answer

Mathematical Olympiads for Elementary School – Problem Book

Problem 5	**Answer**
Problem 6	**Answer**
Problem 7	**Answer**
Problem 8	**Answer**

Exam 6

Problem 1. Add the appropriate arithmetic signs (×, +) to get the correct equality:

$$2\ 8\ 2 = 2\ 0\ 1\ 6$$

Problem 2. Several rings were thrown on the table. Some of them are intertwined with each other. What rings must be cut to divide the structure into separate rings? Cut as few rings as possible.

Problem 3. Five friends, OH, AH, EH, AY and OY, went fishing. Each caught exactly 1 fish. Two caught perch and three caught crucian carp. Furthermore, OY and OH caught different fish, OH and AH also caught different fish, AH and AY as well as AY and EH caught different fish as well. Who caught what fish?

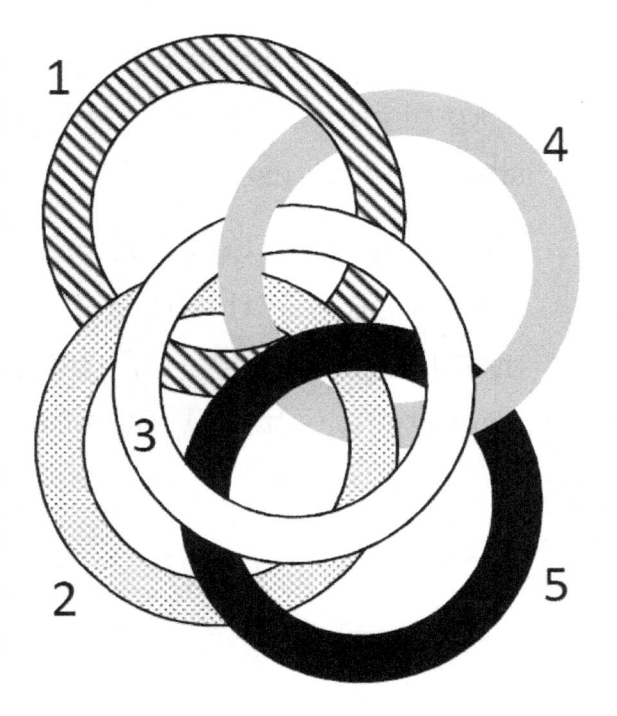

Problem 4. Several owners and their pets met at a cat and dog show. Smart dog Sonya counted that there were 6 heads and 20 paws in total. How many cats would there be if there were more than dogs? (Sonya counted herself, counted her paws with everyone else's)

Problem 5. Cut the shape of the figure into two identical ones.

Problem 6. Dunno and Rustyayka were hastily preparing to travel. Dunno put on first a T-shirt, then a sweatshirt, then a shirt, then a vest.

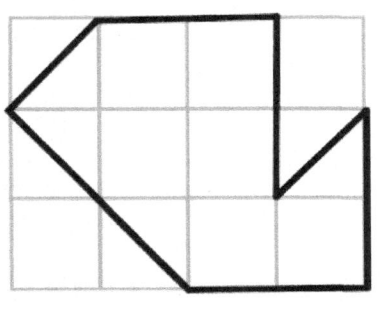

Rustyayka donned the same clothing, but did not wear a single item of clothing in the same order as Dunno. And the garments that are neighboring in Dunno, in Rustyayka are not. In what order did Rustyayka dress if Dunno put on a shirt first?

Problem 7. The two greedy cubs have two pieces of cheese: 4 *kg* and 8 *kg*. They want to share the cheese equally. A fox knows how to divide any piece of cheese into two equal parts, but to do this, then it has to eat 1 *kg* of cheese from any piece that the cubs indicate. How could the cubs share the cheese and give the fox as little as possible?

Problem 8. There are five lamps in a room. Petya said, "There is a lighted lamp in this room". Vasya replied, "You are wrong. There is an unlit lamp in this room". It turned out that of the three statements made, only one was true. Which one?

ANSWER SHEET – EXAM 6

Problem 1	Answer

Problem 2	Answer

Problem 3	Answer

Problem 4	Answer

Problem 5	Answer

Problem 6	Answer

Problem 7	Answer

Problem 8	Answer

Problems – Level 2

Exam 7

Problem 1. There are 6 elephants in a row in increasing order of their height. Which two elephants should be swapped so that there are not three elephants lined up in increasing order at their height?

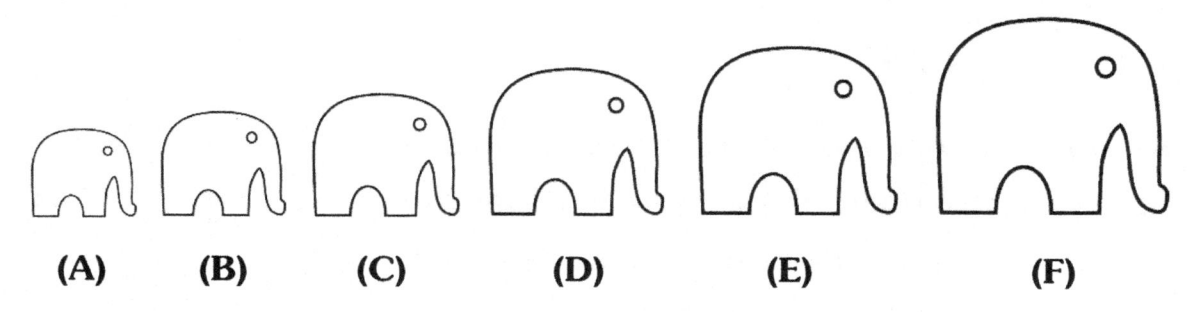

(A) (B) (C) (D) (E) (F)

Problem 2. Put the numbers 1, 2, 3, 4 in the cells so that the 4 numbers are present in each row and in each column, if the numbers 2 and 4 can only be in gray cells.

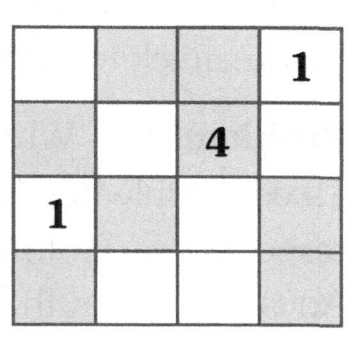

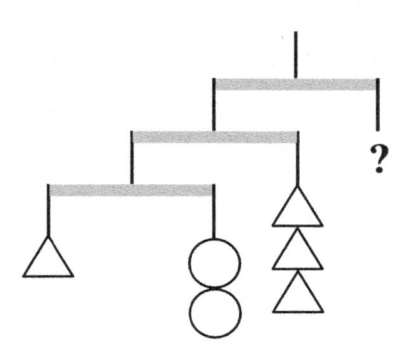

Problem 3. The design of the figure is balanced. All identical figures weigh the same. If the threads are weightless. How many circles must hang in place of the question mark to balance?

Problem 4. Anya, Borya, Vitya, Galya, Dasha, and Zhenya formed a circle. It should always be counted in a clockwise direction. If we start counting from Anya, then Borya will be fifth, if we start from Vitya, Galya will be third, and if we start counting from Zhenya, then Dasha will be fourth. In what order are the children?

Problem 5. Cut the shape shown in the figure into 2 equal parts.

Mathematical Olympiads for Elementary School – *Problem Book*

Problem 6. There are two cuckoo clocks hanging in Grandpa's room. The cuckoos sound every hour as many times according to the number of hours completed and once more every half hour (for example, at 6:00 the cuckoo will sound 6 times and at 6:30 it will sound 1 time). If one of the clocks shows the exact time, and the second is 15 minutes late. How many "cuckoos" will grandson Grisha hear while visiting his grandfather from 12:00 to 3:20?

Problem 7. Petya and Vasya played the naval battle on a 7 × 7 board. Vasya painted over the cells where Petya no longer has ships. Petya still has a three-deck ship ⊏⊐. Which 2 cells should Vasya attack to safely hit Petya's ship?

Problem 8. "Who broke the cup?" – Mom asked sternly to Anya, Vanya and Petya. In response, each of them pointed to one of the other two. It is known that, Anya told the truth. If each child pointed not to whom he pointed, but to another, then the only one who would tell the truth would be Petya. So who broke the cup?

Problems – Level 2

ANSWER SHEET – EXAM 7

Problem 1	Answer
Problem 2	Answer
Problem 3	Answer
Problem 4	Answer

Problem 5	**Answer**
Problem 6	**Answer**
Problem 7	**Answer**
Problem 8	**Answer**

Problems – Level 2

Exam 8

Problem 1. My friends are a cook and a doctor. The doctor's father is Nikolai Petrovich and the cook's father is Ivan Vasilyevich. What is the profession of one of my friends whose grandfather is Pyotr Ivanovich who had no daughters?

Problem 2. Petya assembled a cube of 27 small cubes and then divided it into two parts. A part is shown in the figure to the right. In which figure is the second part of Petya's cube shown?

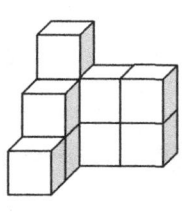

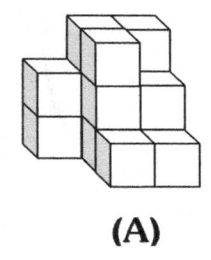

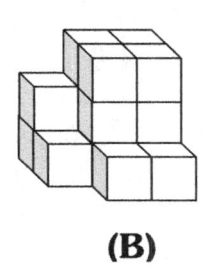

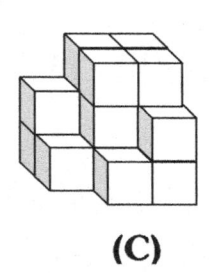

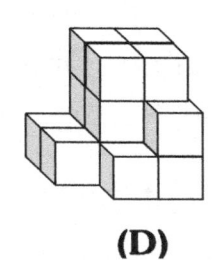

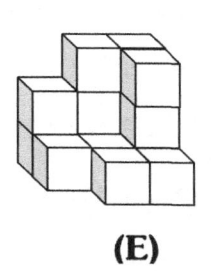

(A)   (B)   (C)   (D)   (E)

Problem 3. Cheburashka learns to add with dominoes. For example, ▨ + ▨ = ▨ means 42 + 13 = 55. Use the dominoes ▨, ▨ and ▨ to make a correct example of adding two-digit numbers.

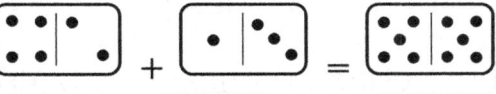

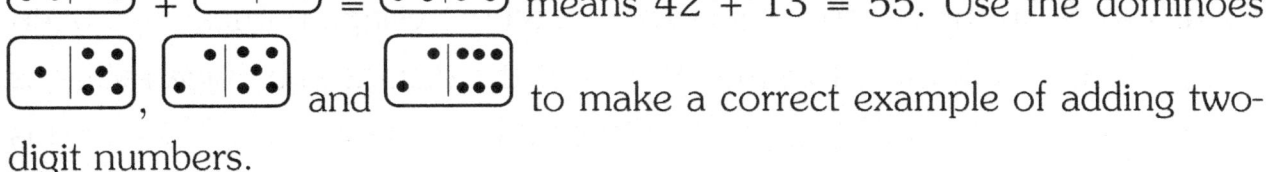

Problem 4. Vintik and Shpuntik get on the Ferris wheel. Vintik is in cabin # 7, and Shpuntik is in cabin # 29. When Shpuntik was at the highest point, there were 3 cabins between Vintik and the lowest booth. How many cabins can there be on the Ferris wheel if the distances between adjacent cabins are the same everywhere? (Check all possible options)

145

Mathematical Olympiads for Elementary School – *Problem Book*

Problem 5. Snezhana forms numbers with matches. She places the number of this year (2018) in front of a mirror in every possible way, but without turning it. What other four-digit numbers can be obtained this way?

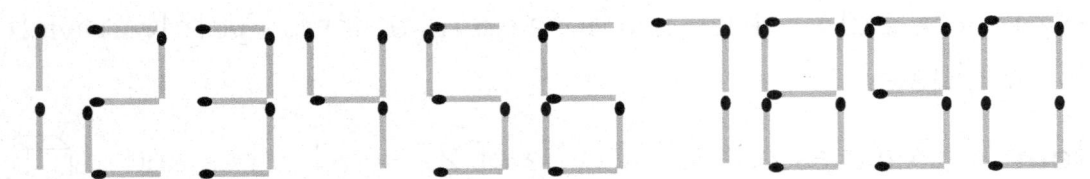

Problem 6. One day, Rabbit's wall clock fell and shattered. The dial was divided into three pieces. Rabbit realized that the sum of the digits of all the pieces was the same. Draw how the dial could have been broken.

Problem 7. One of the brothers Vasya or Sasha ate all the sweets. To the mother's question "Who did it?" Vasya said, "It was the oldest". Sasha said, "It wasn't me". It is known that the one who ate the sweets lied. Who is the oldest?

Problem 8. In the distance table shown, all the village names have been deleted except Okhovo. But the lengths of the distances between villages have been preserved. Retrieve the rest of the village names. (if the cell is empty, it means there is no direct path between the villages)

	?	?	?	?	?	O
?		2	5	8	6	3
?	2		1			
?	5	1		4		
?	8		4		4	6
?	6			4		3
O	3			6	3	

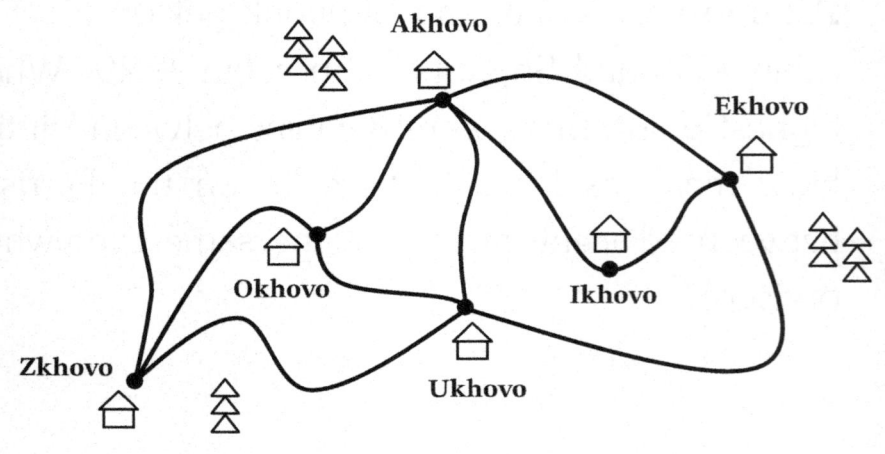

146

Problems – Level 2

ANSWER SHEET – EXAM 8

Problem 1	Answer
Problem 2	Answer
Problem 3	Answer
Problem 4	Answer

Mathematical Olympiads for Elementary School – *Problem Book*

Problem 5	Answer
Problem 6	Answer
Problem 7	Answer
Problem 8	Answer

Problems – Level 2

Exam 9

Problem 1. Replace the letters with numbers from 1 to 7 (If different letters represent different numbers) to get the correct equalities:

$$O \times L = I + M = P + I + A = D \times A$$

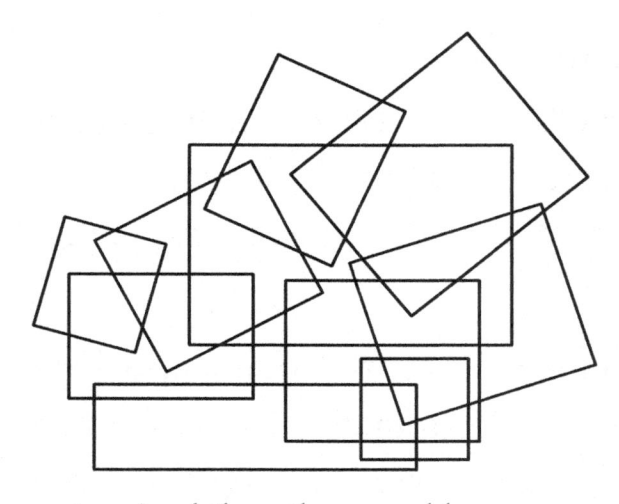

Problem 2. There were ten sheets of paper on the floor. Mike threw three darts and all the sheets of paper were pinned to the floor. Where did Mike throw the darts?

Problem 3. Karabas-Barabas weighed coins. It turned out that the silver coin is heavier than the gold coin. And the silver and bronze ones weigh the same as two gold ones. List the coins in descending order of weight.

Problem 4. Pierrot went to visit his friend Malvina. After leaving the house, he turned right, walked along 3 houses (not counting his own house), turned right again, walked along 2 houses, turned left, went through a house, turned Again to the left, then he passed 3 more houses and in front to the right he saw Malvina's house. In which house does Pierrot live and in which does Malvina live?

Problem 5. Cut the shape of the figure along the triangular edges into three equal pieces.

Problem 6. In the figure shown, place the numbers 1, 2, 3, 4 in the small triangles so that in any triangle of the shape △ or ▽ there are all four numbers.

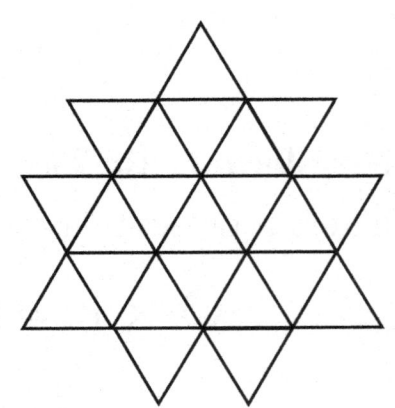

Problem 7. Winnie the Pooh wrote a two-digit number on a piece of paper. The wise owl, looking at the paper, said, if you add 3 to this number, you get a two-digit number, but if you add 9, then you get a three-digit number. Likewise, Owl noted that dividing the number of tens by the number of ones results in a single-digit number with no remainder. What number did Winnie the Pooh write?

Problem 8. Nikita places in a row on a table: a red triangle, a blue square, a yellow circle, a yellow square, a green triangle. Then take two adjacent pieces and then two pieces of the same shape, leaving one piece on the table. What shape is it?

ANSWER SHEET – EXAM 9

Problem 1	Answer

Problem 2	Answer

Problem 3	Answer

Problem 4	Answer

Problem 5	Answer
Problem 6	Answer
Problem 7	Answer
Problem 8	Answer

Problems – Level 2

Exam 10

Problem 1. There are cards with the numbers 0, 1, 2, 3, 4, 5 and the signs "+", "−" and "=". There are 9 cards in total. Using all the cards, make the correct equality. (A number cannot start at 0 unless it is the same number)

Problem 2. The Emerald City guards reported that they saw the shadow of a witch flying over the city. Who did the guards see if they had pictures of all the witches?

Anfisa   *Bastinda*   *Riana*   *Danida*   *Gingema*   *Violetta*

Problem 3. Vasya received a rectangular cake with three types of garnishes. Vasya wants to cut the cake with two straight cuts into four pieces so that each garnish is different. Help him do it! Decorations cannot be cut.

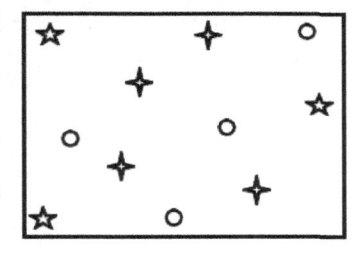

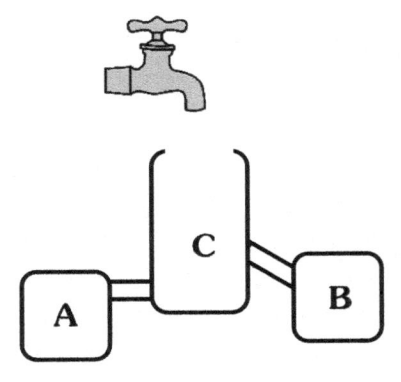

Problem 4. Snusmumrik wants to fill a strange structure with faucet water as in the figure. Which vessel (A, B, or C) will fill first if the faucet is opened?

Problem 5. Baba-Yaga stole Carlson's jam and ran away. After 8 minutes, she noticed the loss and flew in pursuit. At that point Baba Yaga had 10 minutes to flee home, but Carlson runs twice as fast. Will he be in time to catch up with Baba-Yaga?

153

Problem 6. Snow was falling at night. In the morning, Fyodor in boots ⌐⌐, Sharik in sneakers ▩, Matroskin the cat ❀, and Pechkin in felt boots ⌒ walked through the fresh snow. In what order were they?

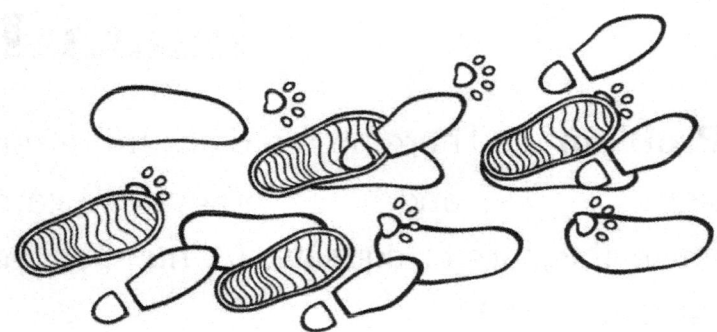

Problem 7. Olya placed three dominoes ⬚⬚⬚ in a rectangle as shown in the figure. It turned out that the sum of points in all vertical rows are equal. Also, the sum of points in the horizontal rows are also the same. Show how Olya arranged the dominoes.

Problem 8. There are four houses along a straight street: blue, yellow, green, and red (in that order). The fox does not live in the red house. And the neighbors of the hare are a bear and a hedgehog. Who lives and where if the fox is not next to the hedgehog?

ANSWER SHEET – EXAM 10

Problem 1	Answer

Problem 2	Answer

Problem 3	Answer

Problem 4	Answer

Mathematical Olympiads for Elementary School – *Problem Book*

Problem 5	**Answer**

Problem 6	**Answer**

Problem 7	**Answer**

Problem 8	**Answer**

Answers: Level 2

Answers – Level 2

Exam 1

1. 35 cherries.
2. Option (B).
3. Vasya Wolf with Zebra Mask, Tosha the Tiger with Hare Mask, Styopa Hare with Tiger Mask, Zoya Zebra with Wolf Mask.
4. 3 green squares.
5. A) 11 points; B) 31 points.
6. Christopher Robin.
7. 20 children.
8. Andrey loves oranges, Vera loves apples.

Exam 2

1. $12 = 34 - 22$.
2. The solution is shown in the figure:

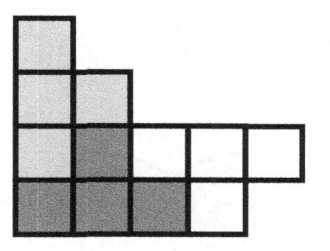

3. the picture of apple number 4 is missing. The figure shows the correspondences of the rest of the apples.

   1 7 2 6 3 5

4. See the table below:

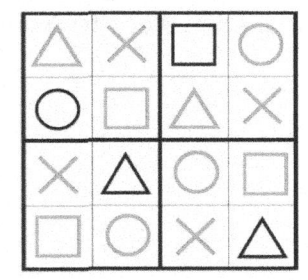

5. The Honda is blue, the Mercedes is yellow, and the Audi is red.
6. 8 kilometers.
7. In 13 minutes.
8. The theoretical physicist..

Answers – Level 2

Exam 3

1. A total of 24 legs.
2. 6 + 2 > 3 < 5 < 6 > 2 + 3 > 4 > 1 + 2.
3. The figure shows one of the possible cut options.

4. There will be 24 hands free.
5. Both Vasya and Julia.
6. The Cheburashka route is: 6 - 2 - 1 - 4 - 6 - 3 - 5 - 4 - 3 - 2.
7. There is 12 russulas in the glade.
8. The longest stick went to Nyusha.

Exam 4

1. One of the options is 2 + 3 + 4 + 5 + 6 + 7 + 8 + 9 + 0 = 44.
2. Rotate counterclockwise.

3. Tune in "AGA" at 12:30 and then tune in "OGO" at 14:30.
4. See the figure below:

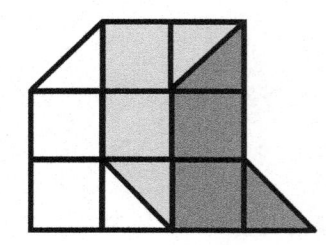

5. The instructor's name is for example Semyon Petrovich Smekhov or Semyon Pavlovich Zadornov.
6. This number is 625143.
7. the carnation is in the white pot, the rose in the yellow one and the lily in the red one. This order can be considered from left to right or from right to left.
8. 10 girls and 7 boys.

Answers – Level 2

Exam 5

1. IVAN.

2. 5 pieces.

3. The number is 16795.

4. An example is shown as follows:

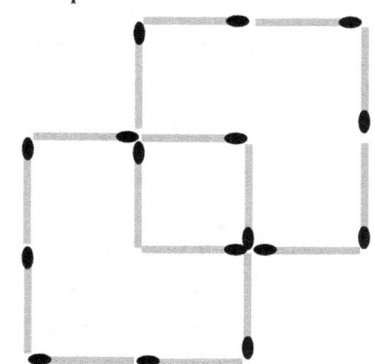

5. He went to bed at 8 pm (clock D) and woke up at 10 am (clock A).

6. 6 toes.

7. 50.

8. Grisha's yellow car arrived first.

Exam 6

1. $2 \times 8 + 2 = 2 + 0 + 16$.

2. You need to cut the gray ring (number 4).

3. OY, AH and EH captured crucian carp, and OH and AY captured perches.

4. 3 cats.

5. See the figure below:

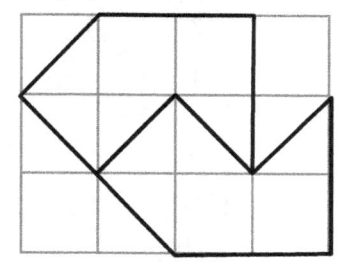

6. Sweatshirt, vest, t-shirt, shirt.

7. The fox will be given the 8 kg piece, it will divide it into two 4 kg pieces and he will eat 1 kg of one. Pieces of 3, 4, 4 kg will then be obtained. Now it will be given a 4 kg piece, he will divide it into two 2 kg pieces and eat 1 kg of one. The resulting pieces will be 1 kg, 2 kg, 3 kg and 4 kg. Then one of the cubs takes the 1 and 4 kg pieces, and the other the 2 and 3 kg pieces.

8. Petya's statement is true: "There is a lighted lamp in this room".

Answers – Level 2

Exam 7

1. Elephants B and E need to be exchanged.

2. The figure below is shown an example:

3	4	2	1
2	1	4	3
1	2	3	4
4	3	1	2

3. 14 circles.

4. Anya - Vitya - Dasha (Zhenya) - Galya - Borya - Zhenya (Dasha).

5. The figure below is shown an example:

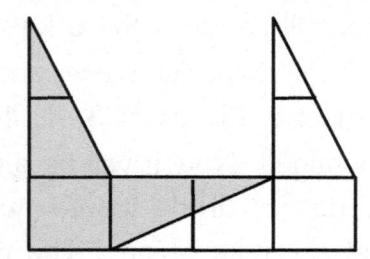

6. 42 cuckoos.

7. The required cells is marked in the figure.

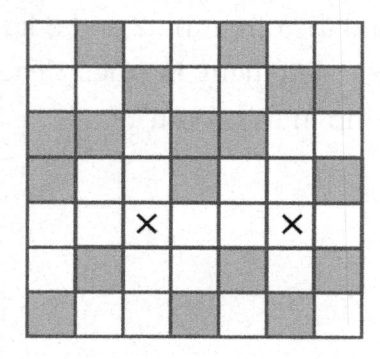

8. Vanya broke the cup.

Exam 8

1. The friend's profession is that of a doctor.

2. Option (B).

3. 25 + 26 = 51.

4. The number of cabins can be 36 or 52.

5. Snezhana will be able to see the numbers 8105, 5018, 3018, 2005, 8118.

6. See the following figure:

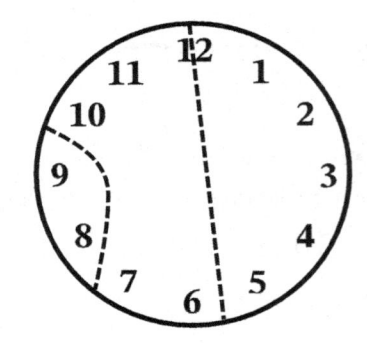

7. The older brother's name is Sasha.

8. The missing info is in the table below:

	A	I	E	U	Z	O
A		2	5	8	6	3
I	2		1			
E	5	1		4		
U	8		4		4	6
Z	6			4		3
O	3			6	3	

Answers – Level 2

Exam 9

1. For example, $2 \times 6 = 7 + 5 = 1 + 7 + 4 = 3 \times 4$.

2. Just throw the darts anywhere in each of the shaded areas.

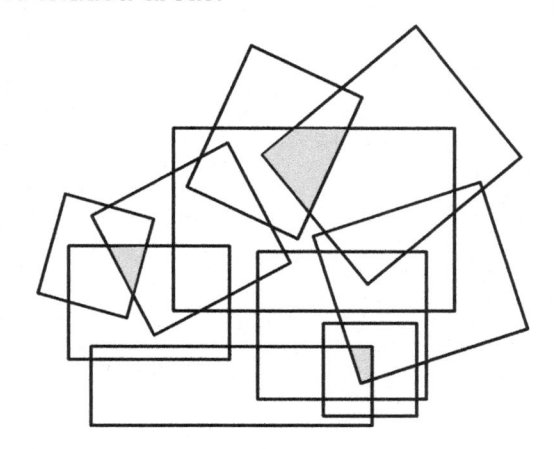

3. In descending order of weights: silver, gold, bronze.

4. There are two options as shown in the figures below:

First option

Second option

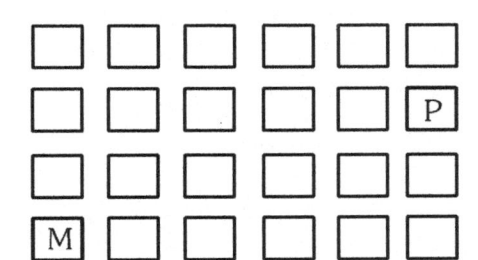

5. An example is shown in the next figure:

6. An option is shown in the figure:

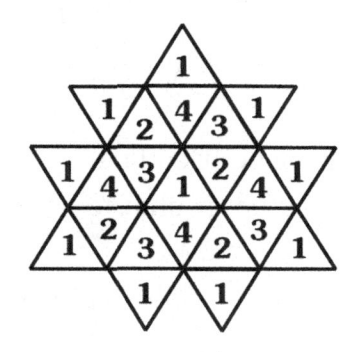

7. Winnie the Pooh wrote 91 or 93.

8. The shape is a square.

Answers – Level 2

Exam 10

1. For example, 30 - 21 = 4 + 5.

2. The guards saw Danida.

3. One of the options is shown in the figure.

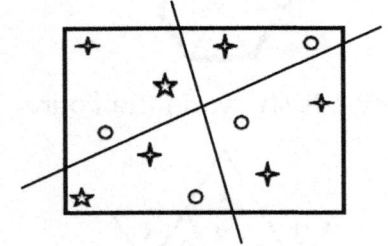

4. Vessel A will be filled first.

5. Carlson will be in time to catch Baba - Yaga.

6. First Pechkin, then Matroskin, then Sharik and last Fedor.

7. An example is shown as follows:

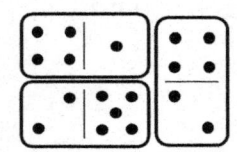

8. The fox is in the blue house, the bear is in the yellow one, the hare is in the green one, and the hedgehog is in the red one.

Problems: Level 3

Exam 1

Problem 1. In a family there are three fathers, three children, a great-grandfather and a great-grandson. How many people are there in all?

Problem 2. Definition 1. "A glove is a wool product in which the fingers are warm".
Definition 2. "A shoe is one that is put on the foot and has laces".
Definition 3. "A sock is a wool product that is worn on the foot".
Definition 4. "A SHOEGLOVE is both a glove and a shoe".
Is the SHOEGLOVE a toe?

Problem 3. Cut the snowflake shown in the figure into 5 pieces with a straight cut.

Problem 4. Kopatych weighs more than Losyash. Hedgehog and Losyash together weigh more than Nyusha and Kopatych together. But Kopatych and Losyash together weigh the same as Hedgehog and Nyusha together. Who weighs more and who weighs less?

Problem 5. Dima has two large coils of rope: white and black. He cuts 10 cm pieces from the coils and ties the three pieces into a 30 cm ring. How many different rings can he get?

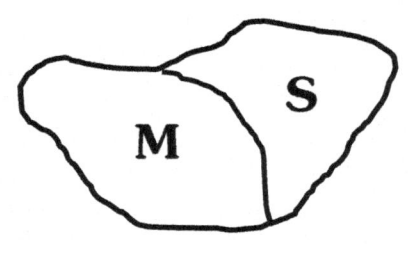

Problem 6. Ranger Stepanych roams the boundaries of his place in 4 hours. While the Ranger Mikhalych walks the boundaries of his place in 6 hours. When Mikhalych retired, his place was incorporated into the Stepanych place

and now Stepanych spends 8 hours roaming the boundaries of the combined places. How long did it take him to travel the shared boundary of these places, if the speeds of the rangers are the same?

Problem 7. Vanya has 8 dominoes (see picture). She wants to arrange them in the shape of a 4 × 4 cell square so that the sum of points in all the rows and all the columns of the square is the same. A) What should this sum be equal to? B) How does Vanya need to arrange the dominoes?

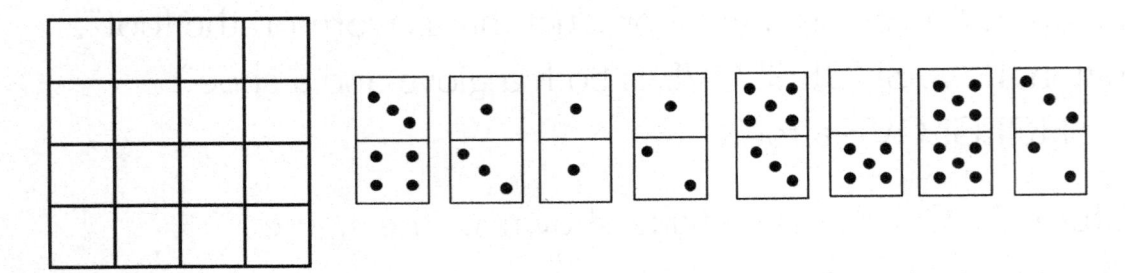

Problem 8. John Silver hid a treasure of gold and silver on three islands: Green Island, Amber Island, and Rocky Island. In one he hid the gold, in another he hid the silver, and in the remaining island he hid nothing. In the bay of each island he posted signs. On the green island: "Gold on the rocky island". On the amber island: "There is no gold or silver here". And on the rocky island: "There is no silver either on the green island or on the amber island". Where is there definitely nothing if all the signs are not telling the truth?

ANSWER SHEET – EXAM 1

Problem 1	Answer

Problem 2	Answer

Problem 3	Answer

Problem 4	Answer

Problem 5	Answer

Problem 6	Answer

Problem 7	Answer

Problem 8	Answer

Problems – Level 3

Exam 2

Problem 1. Between some digits, put an equal sign and an arithmetic sign to get the correct equality:

2 0 0 0 2 0 1 2 1 2

Problem 2. Znayka photographed reflections of the apples in the mirror. If Dunno replaced one picture with another. What apple is Znayka missing the photo of?

1 2 3 4 5 6 7 8

Problem 3. Cut the checkered shape of the figure below into four equal pieces, each with a marked cell.

Problem 4. In the table below, arrange the numbers 1 through 5 so that each column and each row, as well as each highlighted shape, contains all five numbers.

Problem 5. Uncle Fyodor's pace is three times that of Matroskin. First, Matroskin walked a straight path, and then Fyodor, starting from the same place as Matroskin. Following Matroskin's trail, Fyodor clears this trail. So Sharik counted 17 Matroskin footprints. How many footprints of Fedor were on the path?

Problem 6. Winnie the Pooh has 11 large jars of honey and 10 small ones. A store sells boxes in which it can pack 5 large, 9 small, or 4 large and 3 small jars. How many boxes does Winnie have to buy to pack all of his jars of honey? (If he wants to buy as few boxes as possible).

Problem 7. Several Indians and some pale-faced people formed a circle. It is customary for them to lie to their own people and to tell the truth to people with a different skin color. Each turned to his neighbor on the right and said a phrase. If there were 8 phrases "You are an Indian" and 9 – "You are pale-faced". How many Indians and how many pale-faced are there?

Problem 8. Sasha has 2 gold coins, 3 silver and 4 bronze. One of them is false; If the false coin is silver, then it is lighter than real silver; and if the gold or bronze coin is false, then it is heavier than the real gold or bronze coin, respectively. Find the false coin after two weighings on a plate scale. (Note. Coins made of different metals may weigh differently, but real coins made of the same metal weigh the same.)

ANSWER SHEET – EXAM 2

Problem 1	Answer

Problem 2	Answer

Problem 3	Answer

Problem 4	Answer

Mathematical Olympiads for Elementary School – *Problem Book*

Problem 5	Answer

Problem 6	Answer

Problem 7	Answer

Problem 8	Answer

Exam 3

Problem 1. In the next sum, different letters represent different digits. It turned out OLIM + PI + ADA = 2013. Indicate which numbers could be in place of the letters.

Problem 2. My brother always wants to dress differently from me. Therefore, his clothes and shoes are completely different from mine. Determine my brother's name.

YO   PASHA   YASHA   MAXIM   TOLYA   KOLYA   OLEG

Problem 3. In 3rd grade, 6 people eat ice cream every day, 8 people eat ice cream every other day, and the rest eat no ice cream at all. Yesterday 12 students in this class ate ice cream. How many students will eat ice cream today?

Problem 4. In the figure made of matches, five triangles can be counted: four small and one large. Move two matches so that exactly four triangles are visible. There should be no additional matches.

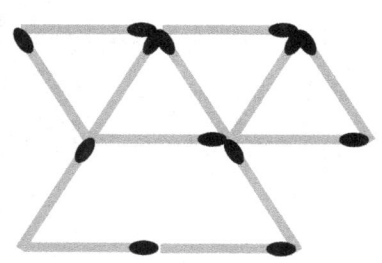

Problem 5. The young physicist Ilya has two identical elastic bands. Likewise, he marked the midpoint of each of them and hung weights on their ends so that one elastic band became twice as long as the other. Ilya measured how far one mark is below the other. How many times is this distance less than the length of the longest elastic band?

Problem 6. Anya, Borya, Vasya and Galya decided to eat a bar of chocolate. But it fell to the ground, and when they picked it up, it turned out that it was broken into five pieces, as shown in the figure. Borya ate the largest piece. Galya and Anya ate the same amount of chocolate, but Galya ate three pieces and Anya one. Vasya ate the rest. What piece of chocolate did Vasya get?

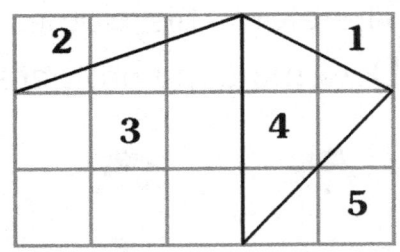

Problem 7. A tinsmith makes letter signs. He makes the same letters at the same time, and different letters, perhaps at different times. If he spent 50 minutes on two signs "DOM MODA" and "DINO", and made the sign "DOMINADO" in 35 minutes. How long does it take to make the sign "ANIDO"?

Problem 8. One day, three friends were talking at a party. Gloria said, "I always speak less than six words". Alex said, "All sentences of more than six words are false". Marty added worriedly, "At least one of us is lying right now". Determine who lied and who told the truth.

ANSWER SHEET – EXAM 3

Problem 1	Answer

Problem 2	Answer

Problem 3	Answer

Problem 4	Answer

Problem 5	Answer
Problem 6	**Answer**
Problem 7	**Answer**
Problem 8	**Answer**

Exam 4

Problem 1. Petya, Vasya, Olya and Masha line up for ice cream. It is known that, the girls are not together, Vasya is right behind Olya and Masha is right behind Petya. What is the order in the line?

Problem 2. Shpuntik was driving a car and saw a kilometer post, on which the number of kilometers was written with a two-digit number with different digits. He drove a little further and saw a kilometer post with the same digits as before, but written in a different order. What is the smallest distance between these posts?

Problem 3. Eight gears are meshed with each other and rotate. Bottom left, clockwise, as shown in the figure. In which direction does the lower right gear rotate?

Problem 4. For the New Year, Santa Claus brought candy for the 3rd graders. If he gives each girl 3 candies and each boy 2, then he will not have enough candies. And if you give each boy 3 candies and each girl give 2 candies, then he has two candies left. What is more in the class: boys or girls, and how much more?

Problem 5. Malish and Carlson each had an identical triangular pyramid made of paper. They cut them out in two different ways. And the left figure shows what Malish got. How will Carlson's cutout be painted (see the right figure)?

Problem 6. Examples of additions were written on a blackboard. Little Johnny replaced different digits with different letters. Resulting that, O + N + E + F + O + R + O + N + E = 32, and S + I + X + F + O + R + S + I + X = 38. What is the next addition: O + N + E + F + O + R + S + I + X?

Problem 7. Popeye "The Sailor" only eats spinach, and exactly once a day: either for breakfast, or for lunch or for dinner. It is known that if Popeye has lunch one day, the next day he will definitely not eat breakfast. If he dines, the next day he will definitely not have breakfast or lunch. If he has only eaten twice in the last two weeks. What meal of the day did Popeye eat his spinach yesterday?

Problem 8. Brothers Avoska and Neboska only lie on their birthdays. Other days, they just tell the truth. Avoska once said: "Today is April 1. Tomorrow is your birthday". Neboska replied, "Today is your birthday. Tomorrow is April 1". When was Avoska born?

Problems – Level 3

ANSWER SHEET – EXAM 4

Problem 1	Answer

Problem 2	Answer

Problem 3	Answer

Problem 4	Answer

Problem 5	Answer

Problem 6	Answer

Problem 7	Answer

Problem 8	Answer

Exam 5

Problem 1. How many three-digit numbers that have the sum of their digits equal to 3 are there?

Problem 2. Kopatych went into hibernation on November 15. But every eighth day funny hares woke him up and he came out of his den. If Kopatych came out of his den ten times. On what date did he leave his den for the last time?

Problem 3. There are 16 boxes on a table. Each one contains an apple or an orange, or the box is empty. It is known that there is an apple and an orange in each row (vertical and horizontal line). Labels are placed around the boxes, indicating which fruit is closest to the edge in this row. Vanya opened one of the boxes and found that it was empty (marked in the figure with a shaded square). Find the fruits that are in the other boxes.

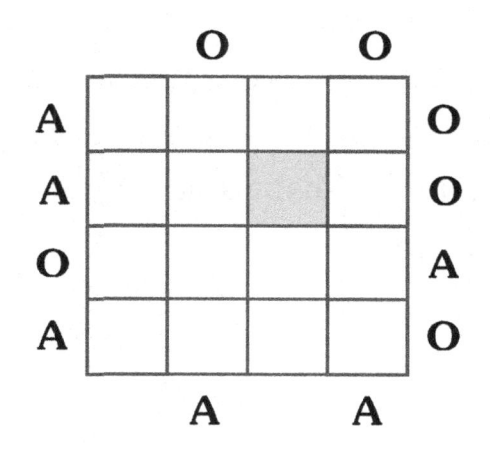

Problem 4. In 2015, Artyom will be 1 year older than the sum of the digits of his year of birth. In what year was Artyom born?

Problem 5. A figure made of matches is shown in the image below. You can see 5 equal squares. Move 4 matches so that only 3 squares can be seen. (There should be no extra matches)

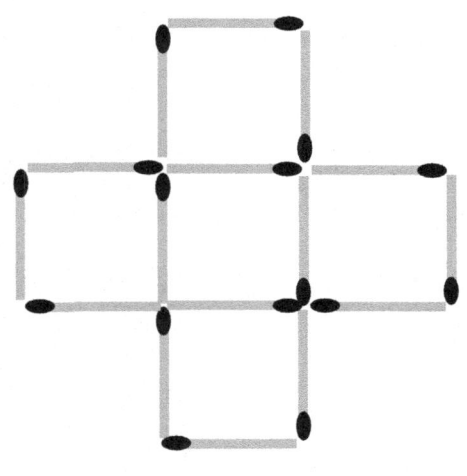

Problem 6. Some children observing a puppy noticed that if it barks, in a minute it eats; if it wags its tail, in a minute it plays; if it

sneezes, in a minute it barks; if it eats, after a minute he wags its tail; if it plays, after a minute it sneezes. The puppy has sneezed now, what will he do in 12 minutes?

Problem 7. Krosh agreed with the Hedgehog to meet in the clearing. However, Krosh's clock is 15 minutes early, but he thinks it is 15 minutes late. And the Hedgehog's clock is 15 minutes late, but he thinks it is 15 minutes early. Who will come to the meeting first and how many minutes will he wait for the other?

Problem 8. There were a few people left in the classroom after the lessons. "Besides me, there are more boys than girls!" - said Nastya. "And besides me, there are more girls than boys!" - said Kolya. "They are both right!" - Misha said. What is the least number of boys and girls left in the classroom?

Problems – Level 3

ANSWER SHEET – EXAM 5

Problem 1	Answer
Problem 2	**Answer**
Problem 3	**Answer**
Problem 4	**Answer**

Problem 5	Answer
Problem 6	**Answer**
Problem 7	**Answer**
Problem 8	**Answer**

Exam 6

Problem 1. Find at least one solution to the numerical puzzle. If different letters correspond to different digits.

$$AAAA + BBB + AA + C = 2016$$

Problem 2. Santa Claus had 5 chocolates with different colored wrappers: red (R), yellow (Y), blue (B), orange (O) and green (G). Five children formed a circle and Santa Claus began to distribute the chocolates in a peculiar way, he skipped a child during the distribution (one yes and the other no). In what order were the chocolates distributed if the last one was in a yellow wrapper, and in the end everyone received chocolates as shown in the figure?

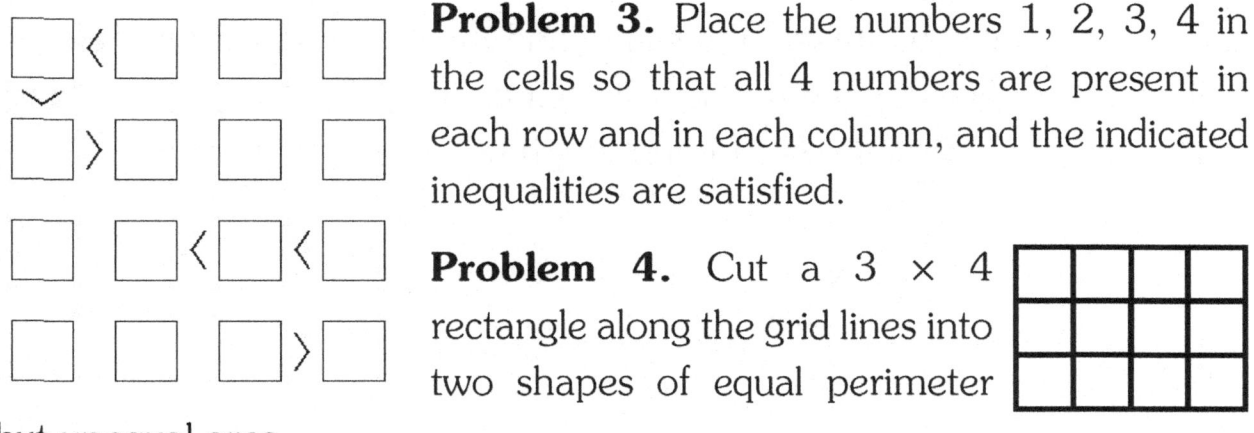

Problem 3. Place the numbers 1, 2, 3, 4 in the cells so that all 4 numbers are present in each row and in each column, and the indicated inequalities are satisfied.

Problem 4. Cut a 3 × 4 rectangle along the grid lines into two shapes of equal perimeter but unequal area.

Problem 5. Usually a child watches 5 cartoons on television before going to bed. But if the Boy is naughty during the day, he is forbidden to see some cartoons. For showing his tongue, the child Freken Bok misses cartoons # 1, 2 and 3. For not eating the meal - cartoons # 2, 4 and 5. For walking on the roof - cartoons # 1 and 5. For jumping over the bed - cartoons # 1 and 4. By playing with the cat - cartoon number 5. In the morning, Freken Bok decided that he wanted to see at least one cartoon today. How many of the listed antics can the child afford on this day?

Problem 6. Edward always says two statements, one of which is true and the other is not. He once said: "Yesterday was Wednesday. The day after tomorrow will be Tuesday". Then he thought a bit and said: "Today is Wednesday. Tuesday was the day before yesterday". What weekday is today?

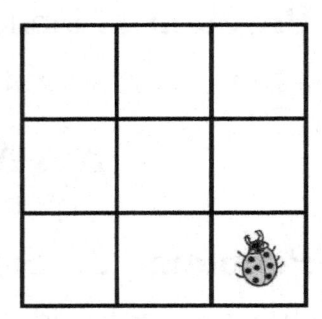

Problem 7. A beetle is in the lower right cell of the board (as in the figure). If he walked through four different cells, and in the fifth he decided to rest. Indicate which cell it could be in, if it is known that the beetle has made an equal number of turns to the left and to the right during its journey.

Problem 8. There are 4 coins in a row on the table. Two of them are known to be fake and weigh the same, and they are also lighter than the real ones. It is also known that counterfeit coins are not neighbors. How to find both counterfeit coins in one weighing with a plate scale?

Problems – Level 3

ANSWER SHEET – EXAM 6

Problem 1	Answer
Problem 2	**Answer**
Problem 3	**Answer**
Problem 4	**Answer**

Problem 5	Answer

Problem 6	Answer

Problem 7	Answer

Problem 8	Answer

Problems – Level 3

Exam 7

Problem 1. Place arithmetic signs to get the correct equality (you can use arithmetic signs and brackets as many times as you like):

1 2 0 2 2 0 1 7 = 5

Problem 2. Given a 7 × 7 squared board. Paint over the cells on this board so that no matter where on the board we choose a cross of five squares (see figure), it covers exactly one painted cell.

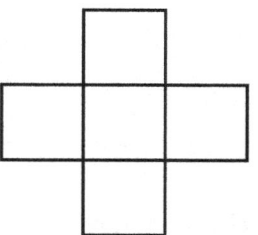

Problem 3. Tikhon represents the natural numbers of a single digit with the help of matches:

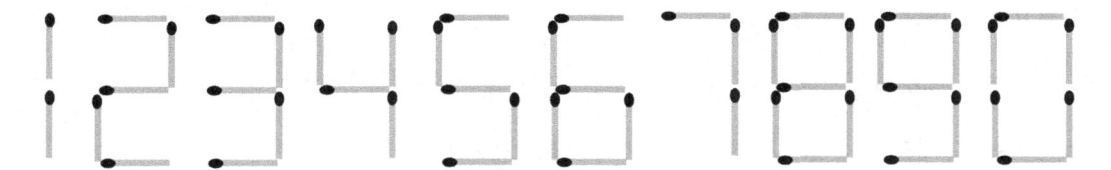

He represents the number "100" as shown in the figure below. Move 4 matches to get as many as possible.

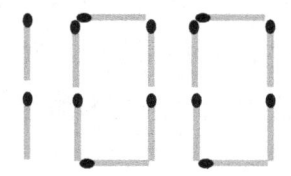

Problem 4. Grisha had three cats: Hasselblad, Vaska and Date. One of them had blue eyes, the other had yellow eyes, and the third had one yellow eye and the other green. If Date had the same eyes as Hasselblad, the total number of eyes of each color would be the same. What eye color does each cat have?

Problem 5. In what order did the pentomino figures fall from top to bottom in the game, if as a result they were positioned as shown in the figure?

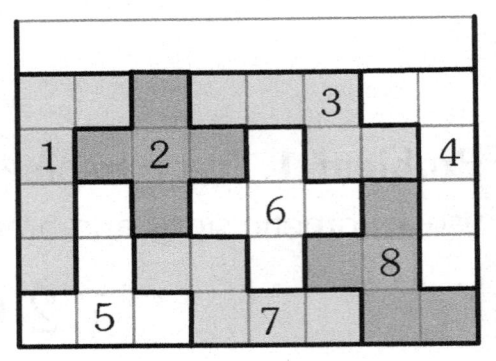

Problem 6. It was decided to number the seats in the single car of a Romashkov train, for which numbered cards were made. It turned out that there were 11 more cards with the number 1 than cards with the number 0. What is the least number of seats in this car?

Problem 7. Cut the Christmas tree in the figure with two straight cuts into multiple pieces so that all the pieces have the same number of balls.

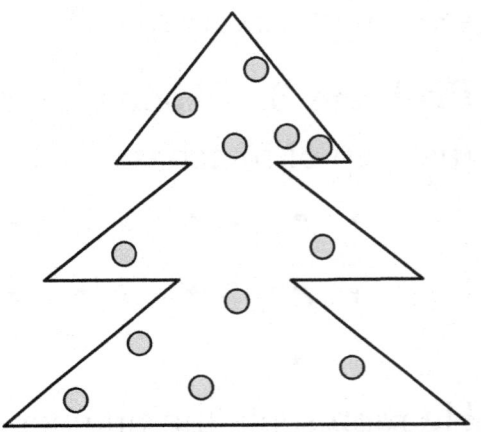

Problem 8. Petya counts the number of apartments in the building where he lives: 1,2,3, ... If the number of apartments is divisible by 11, Petya sneezes, and if the apartment number is divisible by 4, Petya coughs. The floor on which Petya first sneezed and coughed was the second to last. How many floors are there in the building where Petya lives, if there are 4 apartments on each floor of the building (including the first floor)?

Problems – Level 3

ANSWER SHEET – EXAM 7

Problem 1	Answer
Problem 2	**Answer**
Problem 3	**Answer**
Problem 4	**Answer**

Mathematical Olympiads for Elementary School – *Problem Book*

Problem 5	Answer

Problem 6	Answer

Problem 7	Answer

Problem 8	Answer

Problems – Level 3

Exam 8

Problem 1. Karabas-Barabas multiplied three different numbers greater than 1 and got 36. What numbers did Karabas-Barabas multiply?

Problem 2. A vending machine sells three types of chocolates A, B and C. Max wants to buy several chocolates so that he can make a 3 × 3 square with some of them (without breaking them). He notices that there are 1 type A, 3 type B and 7 type C chocolate in the machine. How much will it cost to build this square if each chocolate is worth 10 rubles?

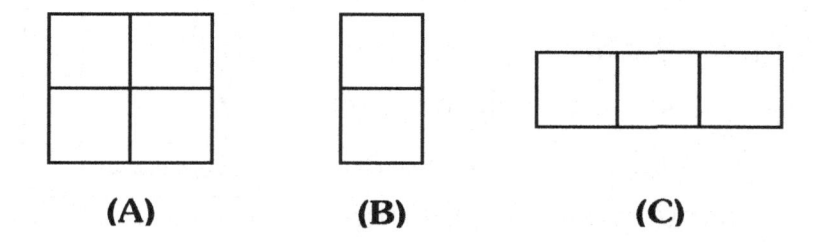

(A)   (B)   (C)

Problem 3. The owners of an art gallery decided to paint the hallway walls in 4 different colors so that the adjacent hallway walls have different colors. Show how they could do it. The gallery plan is shown in the figure below.

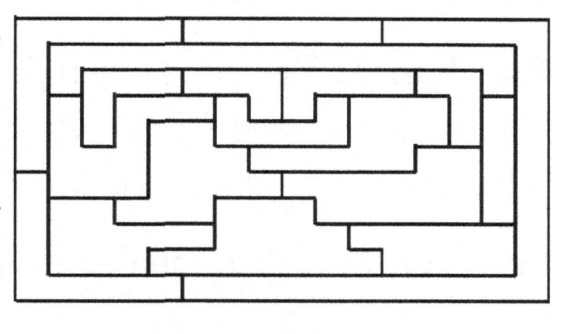

Problem 4. One day, Inga Borisovna's clock fell and shattered. The dial was divided into three pieces. Kolya realized that the sums of the numbers in each of these pieces are three consecutive numbers. Draw how the dial could have been broken.

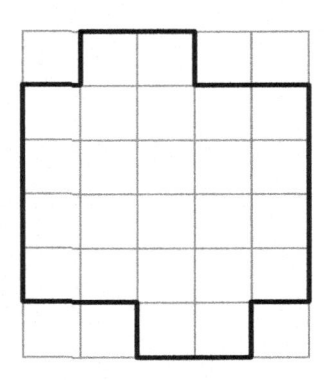

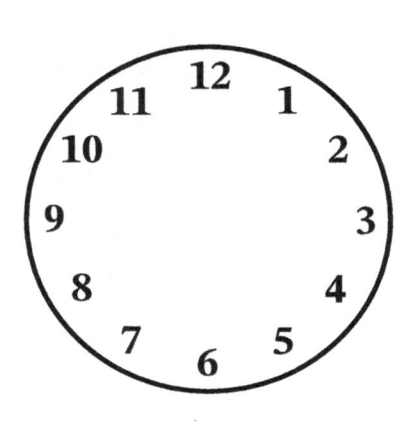

Problem 5. Cut the shape along the grid lines into 4 equal parts. (parts may be rotated)

Problem 6. A swan, a crab and a fish try to move a boat together with the help of ropes for 2 hours. The swan pulls forward for 10 minutes, then 10 minutes backward, then 10 minutes to the left and 10 minutes to the right, again 10 minutes forward, and so on. The crab pulls 15 minutes backward, then 15 minutes to the left, then 15 minutes to the right, backward again, and so on. The fish pulls 20 minutes to the right, 20 minutes to the left, 20 minutes forward, again to the right, and so on. If the boat only moves when everyone is pulling in the same direction. How many minutes did the boat move during these 2 hours?

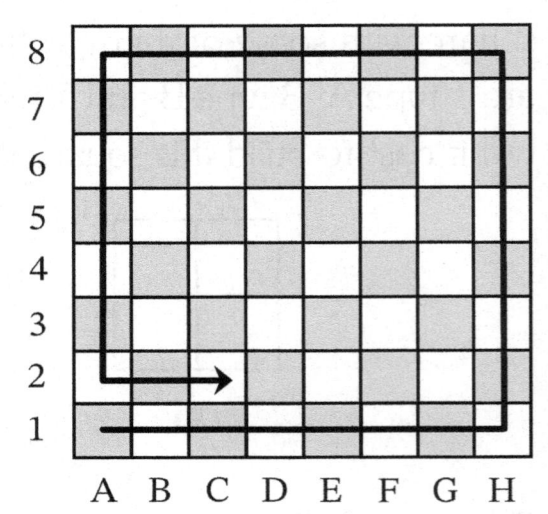

Problem 7. Dominoes with dots from 0 to 6 were placed in a spiral on a chess board (see figure). At one point, all the dominoes were placed. What are the cells that the last domino covered?

Problem 8. Three inhabitants of the island of gentlemen and liars met. One said, "We are all liars". The second objected: "We are all gentlemen!" and the third said nothing. Determine who is who if liars always lie and gentlemen always tell the truth.

Problems – Level 3

ANSWER SHEET – EXAM 8

Problem 1	Answer

Problem 2	Answer

Problem 3	Answer

Problem 4	Answer

Problem 5	Answer
Problem 6	Answer
Problem 7	Answer
Problem 8	Answer

Exam 9

Problem 1. Carlson received a box of sweets. In the morning he ate a third of all sweets, at lunch he ate 2 less sweets than in the morning. And for dinner she finished the remaining 9 sweets. How many sweets were in the box?

Problem 2. Oleg folded a sheet of paper in four, as shown in the image, and made a straight cut. Then he unfolded the sheet. What figure could not turn out?

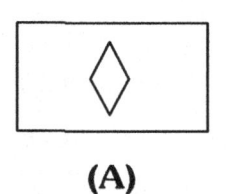

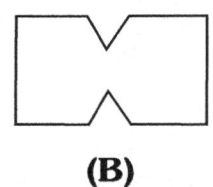

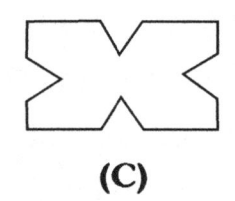

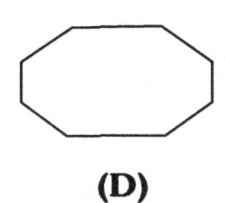

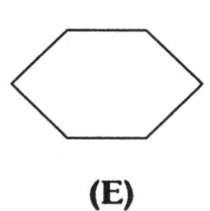

(A)   (B)   (C)   (D)   (E)

Problem 3. Write the numbers 1 through 9 in the small triangles so that in each large triangle the sum of the four numbers is 25.

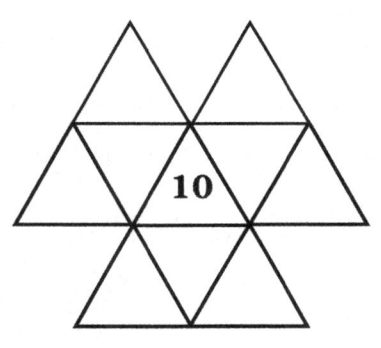

Problem 4. Luke Skywalker took possession of the magic monogram made of gold wire (see figure). To deprive Emperor Palpatine of power, Luke must cut the monogram into exactly 6 pieces with a laser cannon. Luke is tasked with shooting at just 3 points. How can you cope with this task?

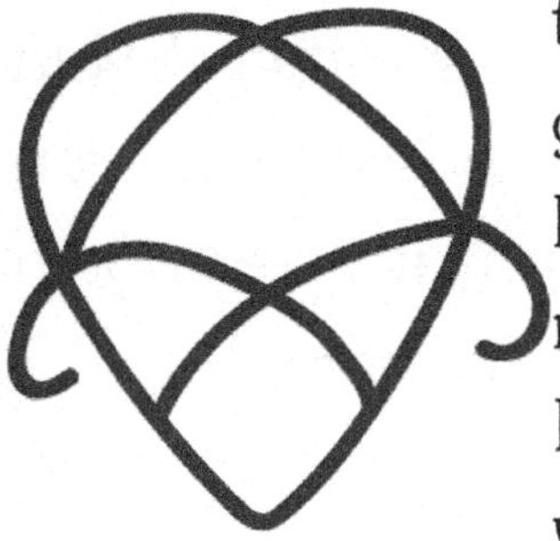

Problem 5. Uncle Fyodor told Sharik a two-digit number. Sharik discovered that if this number is multiplied by 3, you get a two-digit number, and if you subtract 3 from the original number and then divide

the result by 3, you will also get a two-digit number. What number did Uncle Fedor tell Sharik?

Problem 6. At Hogwarts, each of the faculties (Gryffindor, Slytherin, Ravenclaw, and Hufflepuff) has its own library and its own gym. The rules require that each faculty have its own path between the library and the gym, covered with a path of its own color, and the paths must not cross. Draw these 4 paths on the map shown below. (It can only move from room to room through its common side)

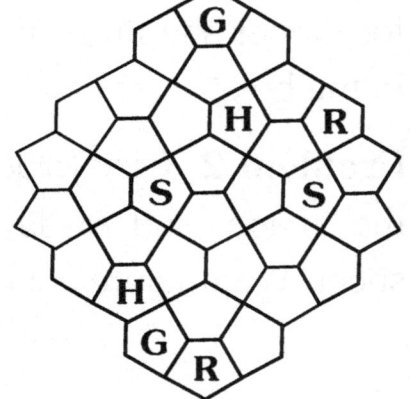

Problem 7. On April 12, *the shorties* launched a rocket to Mars. Znayka told her friends that the rocket on Mars would not appear immediately, but after a while. And to the question "How long?" she silently showed a finger. Friends immediately shouted versions of what Znayka meant: second, minute, hour, day, week, month. To this Znayka replied: "One of you got it right and the others got it wrong, 24, 60, 168, 720, 3600 times". After what time, according to Znayka, will the rocket be on Mars?

Problem 8. Three friends, Masha, Sveta and Dasha, were born the same year, but at different times of the year: winter, spring and summer. Sveta is younger than Dasha and more than six months have passed between Masha and Dasha's birthdays. When was each one born, if it is known that on September 1, they are not all the same number of years?

ANSWER SHEET – EXAM 9

Problem 1	Answer
Problem 2	**Answer**
Problem 3	**Answer**
Problem 4	**Answer**

Problem 5	Answer

Problem 6	Answer

Problem 7	Answer

Problem 8	Answer

Problems – Level 3

Exam 10

Problem 1. Replace different letters with different digits so that you get the correct equality:

$$M + A + N + A + D + A = U \times P \times A$$

Problem 2. Masha (M), Petya (P) and Katya (K) live on the banks of a winding river. Alice got hold of a map of the area where the children live. Could you determine who lives on one bank and who lives on another? If yes, please indicate who lives on the same bank. If there are no other bodies of water in this area.

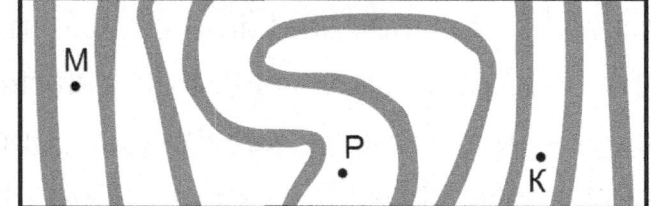

Problem 3. If you pour water into the structure in the figure to the side, vessel A will be filled first. Which vessel will fill first if you open the faucet in the structure in the figure below?

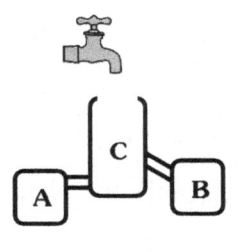

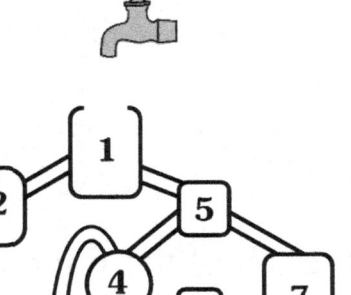

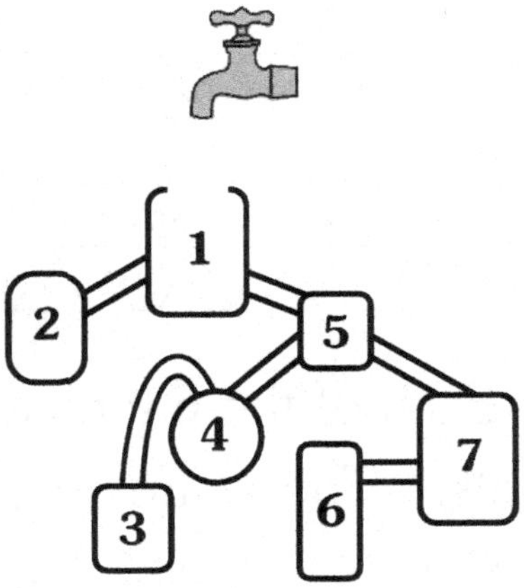

Problem 4. From home to school, Klim has two intersections with traffic lights showing green and red signals. Klim takes 2 minutes to walk from the first to the second traffic light. Klim knows that at each traffic light, the green and red lights are on for the same amount of time, 2 minutes each. It is known that it takes him 10 minutes to walk from home to the first traffic light and 10 minutes from the second traffic light to school. Once Klim left the house at 8:00 and saw that all the traffic lights were simultaneously turning green. What time will he get to school if he does not break the rules? (Klim crosses the street in 5 seconds)

Problem 5. If the smallest three-digit number that is not divisible by 2 is added to the largest three-digit number divisible by 2. What is the sum?

Problem 6. In a slot machine game, one coin must be placed at the same time in three circles connected in a triangle (where one of the circles is the one in the center). In each circle the number of coins placed in it is indicated (in the circle in the center the screen is broken and the number is not visible). Petya started playing when the numbers in the circles were those in the left figure, and finished, when the numbers in the circles were those in the right figure. How many times did Petya play?

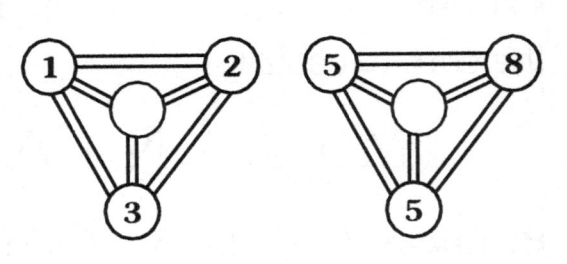

Problem 7. The Hogwarts Museum is divided into triangular rooms. A magic lantern installed in a room illuminates all rooms in three directions (as in the left figure). If a room is lit from three directions, it becomes invisible. Indicate all the invisible rooms in the plan shown in the right figure.

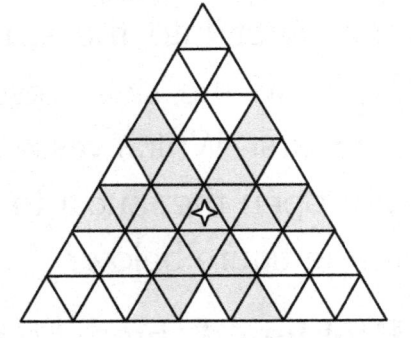

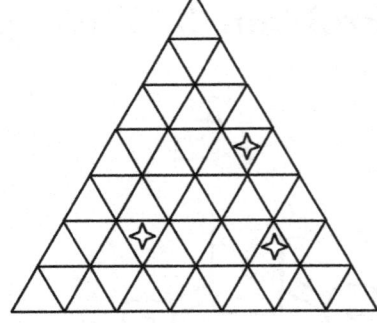

Problem 8. Three friends were playing dominoes. Each of them took a domino from the game and made three statements: "There are four points in my domino"; "My domino has an empty half"; "My domino has the same points in both halves". Which dominoes were taken from the set, if two statements are true and one is false?

Problems – Level 3

ANSWER SHEET – EXAM 10

Problem 1	Answer
Problem 2	**Answer**
Problem 3	**Answer**
Problem 4	**Answer**

Problem 5	Answer

Problem 6	Answer

Problem 7	Answer

Problem 8	Answer

Answers: Level 3

Answers – Level 3

Exam 1

1. 4 people.

2. Yes, it is.

3. An example is shown in the figure:

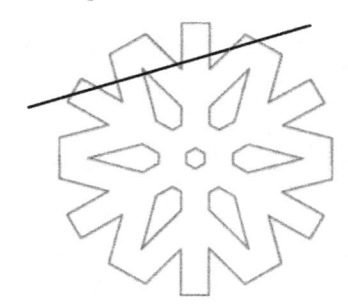

4. The heaviest is the Hedgehog, the lightest is Nyusha.

5. 4 rings.

6. 1 hour.

7. (A) 11; (B) An example is shown in the figure below:

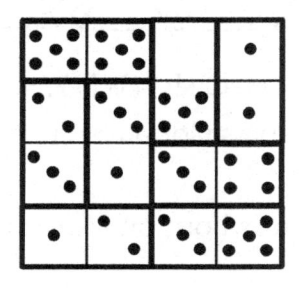

8. on the Rocky Island.

Exam 2

1. 2000 = 2012 - 12.

2. the picture of apple number 3 is missing. The figure shows the matches for the rest of the apples.

6   4   1   5   2   7   8

3. The next figure shows an example.

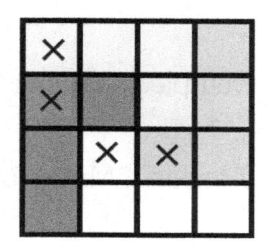

4. The filled table is shown below:

5	3	4	2	1
2	1	5	4	3
1	4	2	3	5
3	2	1	5	4
4	5	3	1	2

5. 9 footprints of Fedor.

6. 3 boxes.

7. 9 Indians and 8 pale-faced.

209

Answers – Level 3

Exam 3

1. For example, 1674 + 57 + 282 = 2013.
2. My brother's name is Maxim.
3. 8 students.
4. An option is shown in the figure. 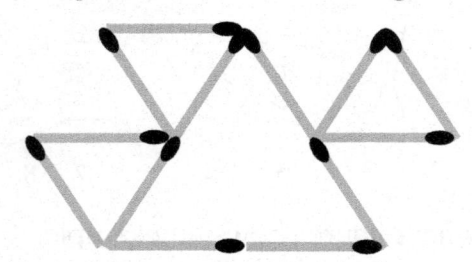
5. 4 times.
6. Vasya received piece number 2.
7. 20 minutes.
8. Gloria and Alex lied. Marty told the truth.

Exam 4

1. Olya, Vasya, Petya, Masha.
2. 9 km.
3. Rotate counterclockwise. 
4. There are 3 more girls in the class.
5. The figure shows all possible options. 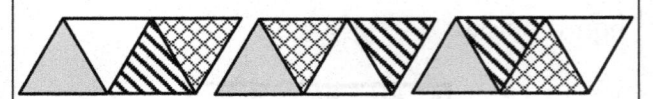
6. O+N+E+F+O+R+S+I+X = 35.
7. Popeye the sailor ate spinach for dinner last night.
8. Avoska was born on March 31.

Answers – Level 3

Exam 5

1. Six numbers. These are the numbers 111, 120, 102, 201, 210 and 300.

2. February 2. Since Kopatych went out only 10 times, by that time 80 days had passed. That means Kopatych came out of his den on the 80th. It will be on February 2nd.

3. The figure below shows the solution.

```
        O     O
  A  [ ][ ][A][O] O
  A  [A][O][ ][ ] O
  O  [O][ ][ ][A] A
  A  [ ][A][O][ ] O
        A     A
```

4. 2006 or 1988.

5. An example is shown as follows:

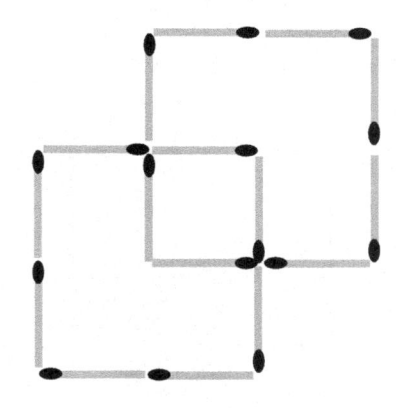

6. It will eat.

7. Krosh will arrive first and wait for Hedgehog for one hour (60 minutes).

8. 2 boys and 2 girls.

Exam 6

1. 1111 + 888 + 11 + 6 = 2016.

2. ROGBY (si se distribuye en sentido antihorario) u ORBGY (si se distribuye en sentido horario).

3. An example is shown below:

3 <	4	1	2
2 >	1	4	3
1	2 <	3 <	4
4	3	2 >	1

4. An option is shown in the figure:

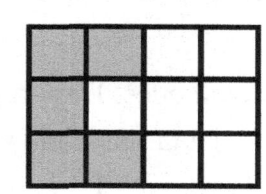

5. The child can afford 4 out of 5 antics.

6. Today is Thursday.

7. The two possible cells are marked with dots in the figure.

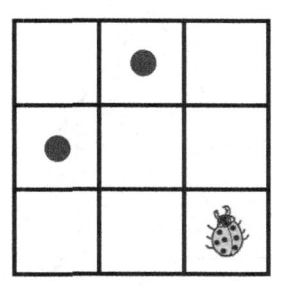

Answers – Level 3

Exam 7

1. An option is: (120: 2 - 20): (1 + 7) = 5

2. The figure below shows an example:

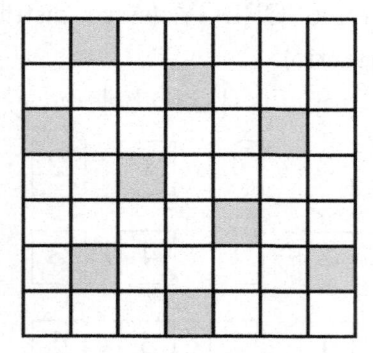

3. The number is 1 111 111

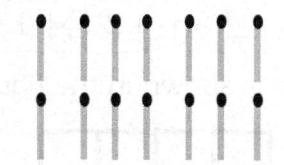

4. Vaska has blue eyes, Date has yellow eyes, and Hasselblad has multi-colored.

5. 5 - 7 - 8 - 6 - 2 - 3 - 4 - 1. For the last three figures there are also options 3 - 1 - 4 and 1 - 3 - 4.

6. 19 seats.

7. An example is shown in the figure.

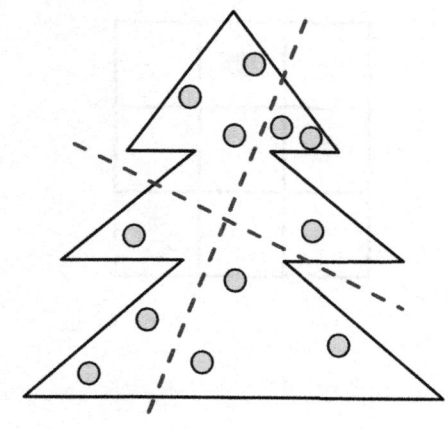

8. 21 floors.

Exam 8

1. The numbers are 2, 3 and 6.

2. 50 rubles.

3. An option is shown in the figure.

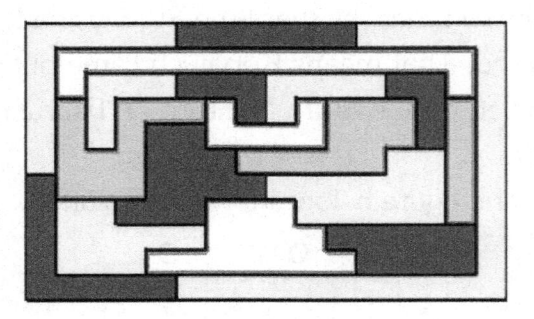

4. An example is shown below:

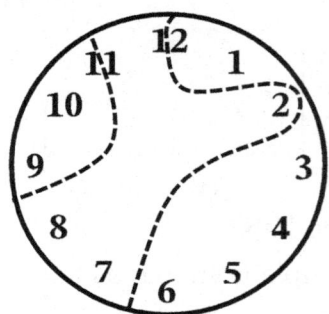

5. An example is as follows:

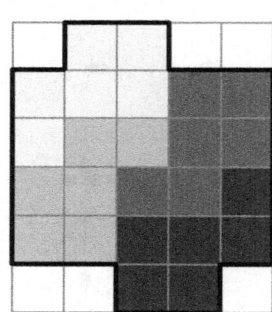

6. 15 minutes.

7. E6-F6.

8. The one who said nothing is a gentleman and the other two are liars.

Answers – Level 3

Exam 9

1. 21 sweets.

2. Option (C). Since to get figure C, you need to make two cuts. Below is how to obtain each of the figures.

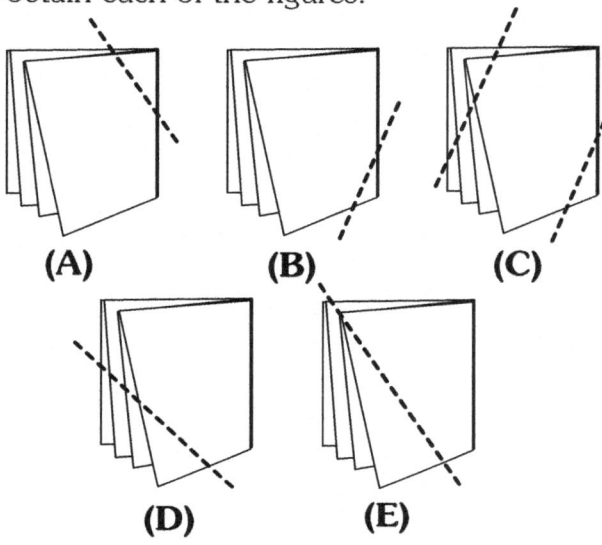

(A) (B) (C)

(D) (E)

3. An example is shown in the figure below:

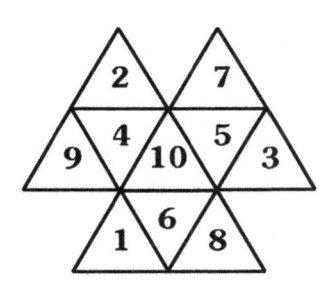

4. An option is shown in the figure below:

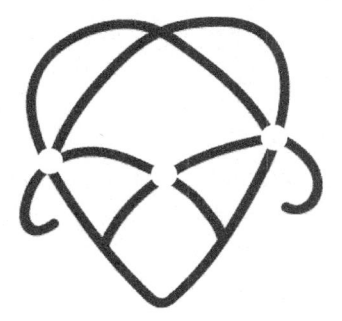

5. The number is 33.

If after multiplying by 3, the two-digit number is still two digits, then this number is not greater than 33. If after dividing by 3 the two-digit number is still two digits, then the number obtained by subtracting 3 to the original number is not less than 30, that is, the original number is not less than 33. Then, this number is at the same time not greater than 33 and not less than 33. Therefore, it can only be 33.

6. An option is shown in the figure:

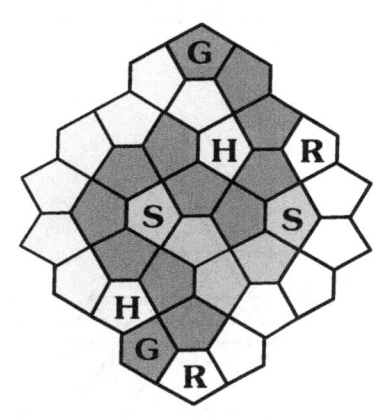

7. An hour later.

8. Dasha - in spring, Sveta - in summer, Masha - in winter, in December.

Answers – Level 3

Exam 10

1. For example, $6 + 1 + 5 + 1 + 0 + 1 = 2 \times 7 \times 1$.

2. Masha and Katya are on one bank; Roma is on the other.

3. Vessel 6 is filled first.

4. Klim will arrive at school at 8:26 and 5 seconds.

5. The sum is 1099.

6. Petya played 6 times.

7. In the figure, the invisible room is shaded gray.

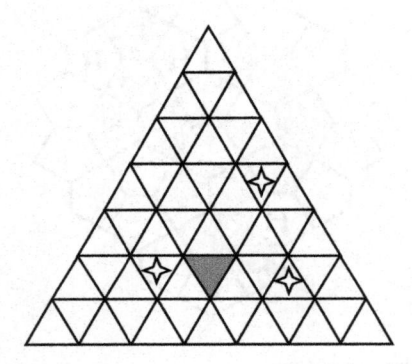

8. 0-0, 4-0 and 2-2. Note that two people cannot have the same true statements. Since two statements define the domino in a unique way. Therefore, three options must be considered and three types of dominoes obtained.

Problems: Level 4

Problems – Level 4

Exam 1

Problem 1. A family consists of four children, three fathers, a grandfather, and two grandchildren. How many people make up the family?

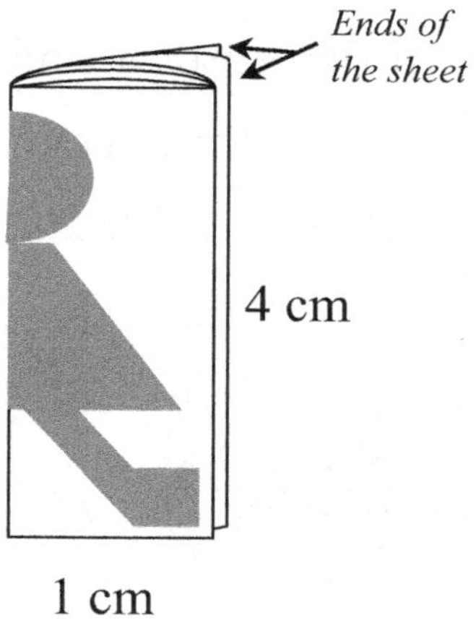

Problem 2. Gosha had a sheet of paper 4 cm wide and 10 cm long. If she folded it several times to make a rectangle 4 cm high and 1 cm wide, and then cut out the figure of a girl. (see picture) How many shapes did he get?

Problem 3. Cut the figure on the left into 4 equal parts.

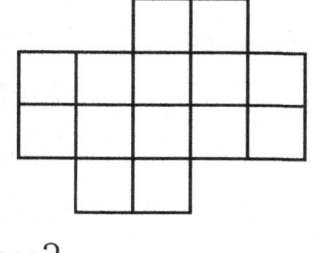

Problem 4. Kopatych weighs more than Losyash. Hedgehog and Losyash together weigh more than Nyusha and Kopatych together. But Kopatych and Losyash together weigh the same as Hedgehog and Nyusha together. Who weighs more and who weighs less?

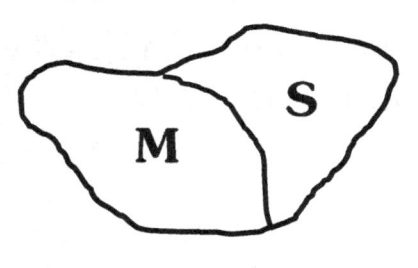

Problem 5. Ranger Stepanych roams the boundaries of his place in 5 hours. While the Ranger Mikhalych walks the boundaries of his place in 6 hours. When Mikhalych retired, his place was incorporated into the Stepanych place and now Stepanych spends 10 hours roaming the boundaries of the combined places. How long did it take him to travel the shared boundary of these places, if the speeds of the rangers are the same?

Problem 6. Dima has three large coils of rope, in blue, red and yellow colors. Cut 10 cm from

the coils and tie three pieces into a 30 cm ring. How many different rings can he get?

Problem 7. Vanya has 8 dominoes (see picture). She wants to arrange them in the shape of a 4 × 4 cell square so that the sum of points in all the rows and all the columns of the square is the same. A) What should this sum be equal to? B) How does Vanya need to arrange the dominoes?

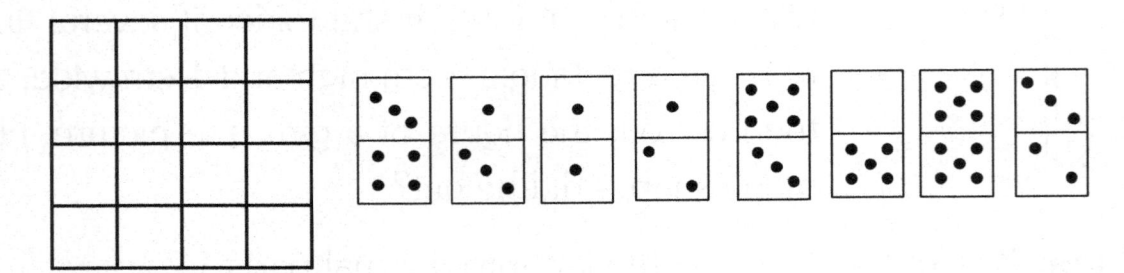

Problem 8. John Silver hid a treasure of gold and silver on three islands: Green Island, Amber Island, and Rocky Island. In one he hid the gold, in another he hid the silver, and in the remaining island he hid nothing. In the bay of each island, he posted signs. On the green island: "Gold on the rocky island". On the amber island: "There is no gold or silver here". And on the rocky island: "There is no silver either on the green island or on the amber island". Where is there definitely nothing if all the signs are not telling the truth?

ANSWER SHEET – EXAM 1

Problem 1	Answer
Problem 2	**Answer**
Problem 3	**Answer**
Problem 4	**Answer**

Problem 5	Answer

Problem 6	Answer

Problem 7	Answer

Problem 8	Answer

Exam 2

Problem 1. Swap the two-digit places to get the correct equality:

$$2012 = 1719 + 275$$

Problem 2. Petya ate a round cake on his birthday, which was sliced down the center. Each slice had a candle, and some slices also had a rose. Masha and Misha began to count the candles in a circle (each one started with a candle), but they both forgot where they started. Masha counted 6 candles and 2 roses, and Misha counted 19 candles and 3 roses. How old is Petya?

Problem 3. Cut the checkered figure into two identical pieces, each of which is an expansion of a $1 \times 1 \times 1$ cube.

Problem 4. In the table below, place the numbers 1 through 7 so that each column and each row, as well as each highlighted figure, contains all seven numbers.

Problem 5. Nikita has a ruler with divisions in centimeters and millimeters. Likewise, Nikita discovered that there are exactly 80-millimeter divisions on the ruler. What is the distance between the first and the last division of Nikita's ruler?

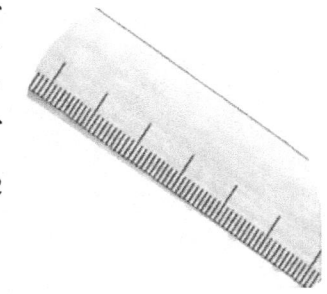

Problem 6. *Winnie the Pooh* has 11 large jars of honey and 10 small ones. A store sells boxes in which it can pack 5 large, 9 small, or 4 large and 3 small jars. How many boxes does Winnie have to buy to pack all of his jars of honey? (If he wants to buy as few boxes as possible).

Problem 7. Andrey, Borya and Vitya came to the Olympiad. One of them is a first grade student, another one is a second grade student, and the remaining one is a third grade student. It is known that the second grader solved one less problem than Andrey and Vitya solved two more problems than the third grader. Who solved more problems and how much more: Borya or a first grader?

Problem 8. Sasha has 2 gold coins, 3 silver and 4 bronze. One of them is false; If the false coin is silver, then it is lighter than real silver; and if the gold or bronze coin is false, then it is heavier than the real gold or bronze coin, respectively. Find the false coin after two weighings on a two-plate scale. (Note. Coins made of different metals may weigh differently, but real coins made of the same metal weigh the same.)

ANSWER SHEET – EXAM 2

Problem 1	Answer

Problem 2	Answer

Problem 3	Answer

Problem 4	Answer

Problem 5	**Answer**
Problem 6	**Answer**
Problem 7	**Answer**
Problem 8	**Answer**

Problems – Level 4

Exam 3

Problem 1. In the next sum, different letters represent different digits. It turned out OLIM + PI + ADA = 2013. Indicate which numbers could be in place of the letters.

Problem 2. I have two friends who hate dressing in the same clothes and shoes. Therefore, they always dress in such a way that everything is different about them. Find these guys among the seven guys shown.

SASHA   NIKITA   YASHA   VASYA   TOLYA   KOLYA   OLEG

Problem 3. The young physicist Ilya has two identical elastic bands. Likewise, he marked the midpoint of each of them and hung weights on their ends so that one elastic band became three times as long as the other. Ilya measured how far one mark is below the other. How many times is this distance less than the length of the longest elastic band?

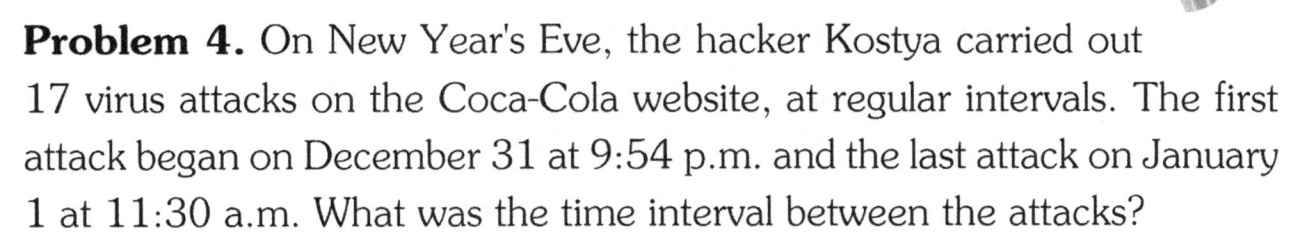

Problem 4. On New Year's Eve, the hacker Kostya carried out 17 virus attacks on the Coca-Cola website, at regular intervals. The first attack began on December 31 at 9:54 p.m. and the last attack on January 1 at 11:30 a.m. What was the time interval between the attacks?

Problem 5. Anya, Borya, Vasya, Galya and Dasha decided to eat a chocolate bar. But this one fell to the ground, and when they picked it up, it turned out that it was broken into seven pieces, as shown in the figure. Borya ate the largest piece. Galya and Dasha ate the same amount of chocolate, but Galya ate three pieces and Dasha one. Vasya ate a seventh of the entire chocolate bar and Anya ate the rest. What piece of chocolate did Anya get?

Problem 6. In the figure shown with matches, six triangles can be counted. Move four of these matches so that nine triangles are visible. Likewise, there should be no extra matches.

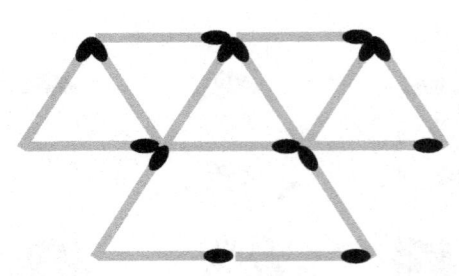

Problem 7. A tinsmith makes letter signs. He makes the same letters at the same time, and different letters, perhaps at different times. If he spent 50 minutes on two signs "DOM MODA" and "DINO", and made the sign "DOMINADO" in 35 minutes. How long does it take to make the sign "ANIDO"?

Problem 8. Once at a party, four friends were talking. Gloria said, "I always speak less than six words". Rico replied: "My statement has no more than eight words!". Alex said, "Gloria and Rico are now telling the truth". Marty added worried: "But today someone, Alex or Gloria, lied". Determine who lied and who told the truth.

ANSWER SHEET – EXAM 3

Problem 1	Answer
Problem 2	**Answer**
Problem 3	**Answer**
Problem 4	**Answer**

Problem 5	Answer
Problem 6	**Answer**
Problem 7	**Answer**
Problem 8	**Answer**

Exam 4

Problem 1. A neighbor's father, son and grandson came to congratulate him on his birthday. His names are Anton Sergeevich, Andrei Borisovich and Sergei Nikitich. What is the name of the neighbor if he has only one son and no daughters?

Problem 2. The postman Pechkin left Prostokvashino and the policeman Svistulkin left Smetanino. They met on a kilometer post, on both sides of which were written the distances to Smetanino and Prostokvashino. Pechkin noticed that these are two different numbers, written with the same numbers, but in a different order. What is the smallest distance between Prostokvashino and Smetanino?

Problem 3. Sharik built a parallelepiped with 2 cm, 4 cm and 6 cm sides. Matroskin built a cube with a 3 cm side. Uncle Fyodor cut a rectangular hole in cardboard, into which Sharik's parallelepiped fits, but Matroskin's cube does not. What size hole could he cut? Just give 1 example.

Problem 4. In the figure made of matches, a small, a medium, and a large triangle are shown. From this figure, construct one in which there are exactly two small, two medium and two large triangles. There should be no additional matches, if each match is to participate in at least one triangle.

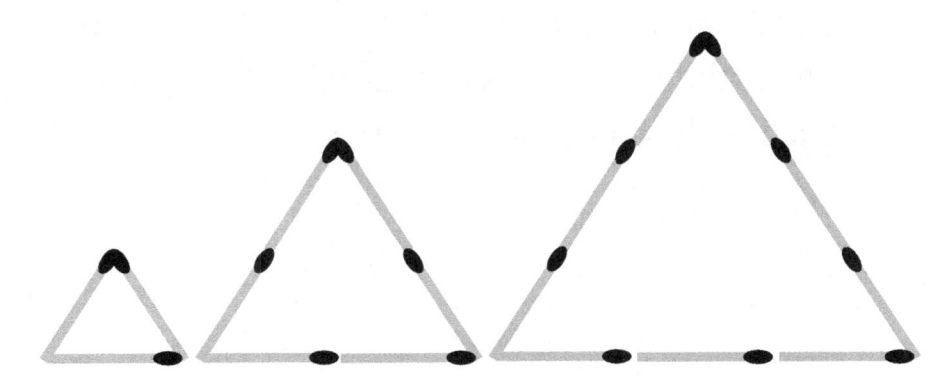

Problem 5. Examples of additions were written on a blackboard. Little Johnny replaced different digits with different letters. It turned that O + N + E + F + O + R + O + N + E = 20, and S + I + X + F + O + R + S + I + X = 50. What is the result of O + N + E + F + O + R + S + I + X?

Problem 6. During some excavations in the territory of Ancient Rome, an unusual clock whose dial had 18 divisions was found, where Roman numerals were used for numbering (see figure). Unfortunately, the watch dial was divided into 5 parts. Nikita noticed that the sums of the numbers in each of the parts are equal. Show how the dial could have been broken.

Problem 7. Popeye "The Sailor" only eats spinach, and exactly once a day: either for breakfast, or for lunch or for dinner. It is known that if Popeye has breakfast on a certain day, the next day he will only have lunch. If he has lunch, the next day he will definitely not have breakfast. And if he dines one day, the next day he will necessarily have breakfast. Popeye had lunch on January 1, and during every day from January 1 to February 8, he ate breakfast as many times as he had dinner. What meal of the day did Popeye eat his spinach on February 8?

Problem 8. Brothers Avoska and Neboska only lie on their birthdays. Other days, they just tell the truth. Avoska once said: "Today is April 1. Tomorrow is your birthday". Neboska replied, "Today is your birthday. Tomorrow is April 1". When was Avoska born?

ANSWER SHEET – EXAM 4

Problem 1	Answer

Problem 2	Answer

Problem 3	Answer

Problem 4	Answer

Problem 5	**Answer**
Problem 6	**Answer**
Problem 7	**Answer**
Problem 8	**Answer**

Exam 5

Problem 1. The Malinkins Masha and Vasya came to visit the Strawnichkins Petya and Sveta. And they brought them food: cookies, cake, chocolates and apples. The girls (Masha and Sveta) ate cookies and apples, and the Malinkins ate cookies and chocolates. They ate all the food; however, Masha does not like apples. What did each eat if everyone ate one of the foods?

Problem 2. Nikita wrote a puzzle about an inequality with two-digit numbers: ON > NO, where different letters represent different digits. How many solutions are there to this puzzle?

Problem 3. Yegor has chocolates and candies. If his mother gives him 10 more candies, the number of candies will be twice as many as the chocolates. Yegor wondered how many chocolates he should give Vitalik so that the remaining sweets also contain twice as many candies as there are chocolates. Help Yegor solve this problem.

Problem 4. In 2015, Nikita will be so old that his age will be equal to the sum of the digits of her year of birth. What year was Nikita born? Find all the possibilities.

Problem 5. An arrangement with matches is shown in the figure below. Also, 3 rhombuses and 4 triangles can be seen. Move 4 matches so that only 1 rhombus and 6 triangles can be seen. (There should be no extra matches)

Problem 6. A bar of chocolate consists of 12 squares of black chocolate and 12 squares of white chocolate, as shown in the figure. Carlson wants to cut a 2 × 2 square so that there are equal parts of black and white chocolate. How many ways can he do it?

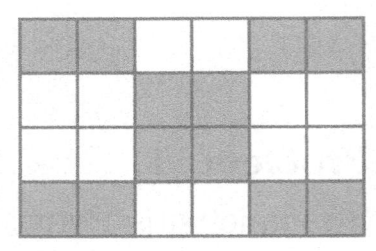

Problem 7. Some children observing a magician and his magic hat noticed that if he puts a handkerchief in his hat, after a minute he takes out a ball; if he puts a ball in his hat, after a minute he takes out a rabbit; if he takes something out of his hat, after a minute he waves his wand; if he waves his wand, in a minute he takes a handkerchief out of his pocket; if he takes a handkerchief out of his pocket, in a minute he puts it in his hat; if he takes a rabbit out of his hat, in a minute he takes a ball out of his pocket; if he takes a ball out of his pocket, in a minute he puts it in his hat. Now the magician took a ball out of his pocket, what will he do in 7 minutes?

Problem 8. A family with three children lived on this island. One day they got together and said among themselves; Sasha: "I have two sisters". Zhenya: "And I also have two sisters". Valya: "And I have two brothers". How many boys and how many girls are there in this family? (Note: the names Sasha, Zhenya and Valya can be used by both boys and girls)

ANSWER SHEET – EXAM 5

Problem 1	Answer
Problem 2	**Answer**
Problem 3	**Answer**
Problem 4	**Answer**

Mathematical Olympiads for Elementary School – *Problem Book*

Problem 5	**Answer**
Problem 6	**Answer**
Problem 7	**Answer**
Problem 8	**Answer**

Exam 6

Problem 1. Three brothers have a different number of coins. If the oldest gave 1 coin to the middle and the middle gave 3 coins to the youngest, then they would all have the same amount. How many coins must the oldest to the youngest of them give so that both have the same?

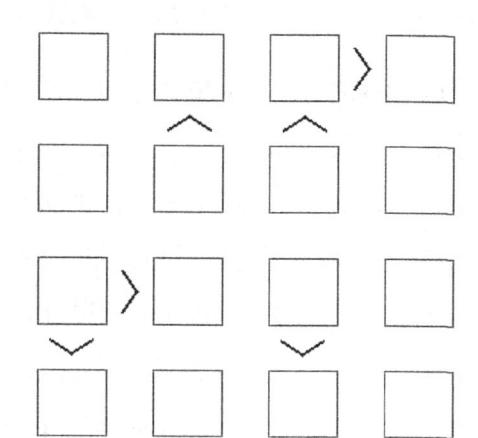

Problem 2. Place the numbers 1, 2, 3, 4 in the cells so that all 4 numbers are present in each row and in each column, and the indicated inequalities are satisfied.

Problem 3. Cut a 6 × 6 square board into two equal 24-sided polygons.

Problem 4. Ekaterina Mikhailovna (EM) had a collection of 10 notebooks with covers of different colors: Red (R), White (W), Black (K), Yellow (Y), Blue (B), Purple (P), Orange (O), Light Blue (L), fuchsia (F) and Green (G). Ten children formed a circle and EM began to distribute the notebooks to one in three children, counting in a circle and skipping those to whom she had already given. In what order were the notebooks, if Yegor received a yellow notebook, and he was the third to receive the notebook from her? The figure shows which notebooks all received in the end.

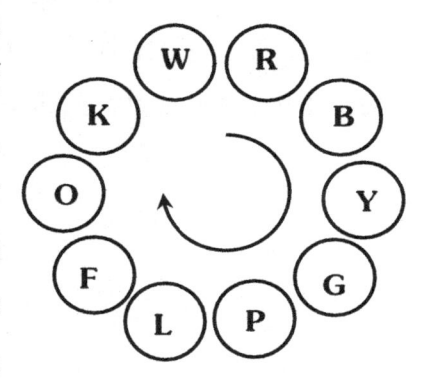

Problem 5. Find at least one solution to the puzzle. (If different letters correspond to different digits)

Problem 6. Masha and Sasha brought an identical packet of cookies to school and agreed to eat 2 or 3 of them at

```
   W I N E
 ×       6
  ---------
   G A I N
```

each break. If at the end of the fourth lesson, Sasha had only one cookie left; and by the sixth lesson, Masha had already run out of cookies. How many cookies were in the packet?

Problem 7. Once, a prince met with three witches to ask them about their fate and they answered him. Arta: The prince will have a lazy wife, and will defeat more than 100 dragons. Binah: No, no, the prince will defeat less than 100 dragons, but the wife will be a hard worker. Veda: No, the wife, alas, will be lazy, but the prince will defeat at least one dragon. What awaits the prince if it is known that one of them always lies, another always tells the truth, and the other one tells the truth first and then lies?

Problem 8. There are 9 coins in a row, it is known that among them there are exactly three that are false, and they are always together. All counterfeit coins weigh the same and are lighter than real coins. And all real coins weigh the same. How to find the three counterfeit coins in 2 weighings on a two-plate scale?

ANSWER SHEET – EXAM 6

Problem 1	Answer
Problem 2	**Answer**
Problem 3	**Answer**
Problem 4	**Answer**

Problem 5	Answer
Problem 6	Answer
Problem 7	Answer
Problem 8	Answer

Exam 7

Problem 1. There are 9 cards with numbers from 1 to 9. Arrange them in a row so that there are not three consecutive cards with numbers in ascending or descending order.

Problem 2. Place the letters A, B and C in the circles so that there are no equilateral triangles with three identical letters at their vertices.

Problem 3. The city has 8 train stations: Alf, Beta, Hammilton, Dhelta, Lambda, Epsilon, Eks and Zeta. It is known that a train runs directly between two stations if the number of letters in the names of these stations has different parities. Fedya wants to travel as long as possible without visiting any station twice, so that the name of each subsequent station is longer than the previous one. How long (stations) will this journey take? Explain your answer.

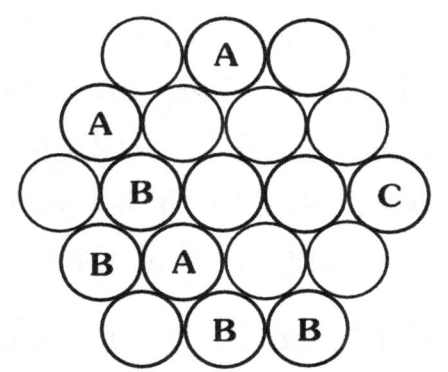

Problem 4. The planet Zhelezyaka makes one revolution around its axis in 5 hours Zhelezyaka. And the planet Kamenyuk makes one revolution around its axis in 6 hours Kamenyuk. A spaceship travels from the planet Zhelezyaka to the planet Kamenyuk taking 20 hours Zhelezyaka, and returns in 25 hours Kamenyuk. Which planet rotates on its axis the fastest? Explain your answer.

Problem 5. It was decided to number the houses along the only street in the City of Flowers, for which plates with numbers were used. It turned out that there were 12 more plates with the number 1 than plates with the number 0. What is the least number of houses on this street? Explain your answer.

Problem 6. Cut the tree in the figure into 13 pieces with just three straight cuts.

Problem 7. Valya, Sasha, Zhenya and Slava played at recess. One of them broke a window, and each of them said; Valya: "One of the men broke it"; Sasha: "It was Slava!"; Zhenya: "There are more men among us"; Slava: "Valya and I are women!". It turned out that all the girls had lied and all the boys had told the truth. Who broke the window? (If all names can be both boys and girls) Explain your answer.

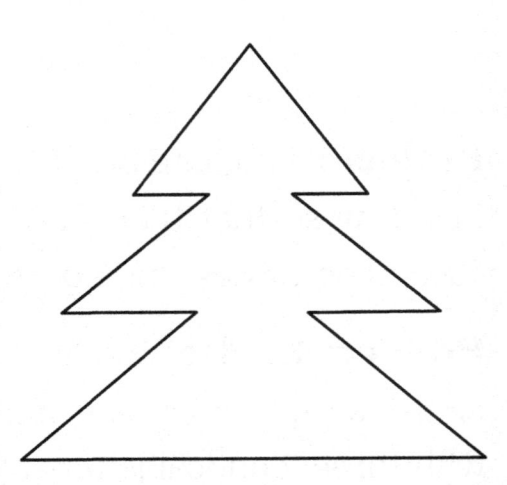

Problem 8. It is required to put an odd number of candies in boxes of 46 units, but only 43 boxes were filled. Then, it was tried to put them in boxes of 43 units, but 47 boxes were filled and there was also something left. Will it be possible to arrange the candies in exactly 17 boxes? Explain your answer.

ANSWER SHEET – EXAM 7

Problem 1	Answer

Problem 2	Answer

Problem 3	Answer

Problem 4	Answer

Problem 5	Answer
Problem 6	Answer
Problem 7	Answer
Problem 8	Answer

Exam 8

Problem 1. A cake costs the same as two turnovers and three turnovers cost as much as two chocolates. What is more expensive, two cakes or three chocolates?

Problem 2. We can make numbers with the dominoes. For example, ▨▨ represents the number 4203. Perform the correct operation of adding two four-digit numbers with the following dominoes:

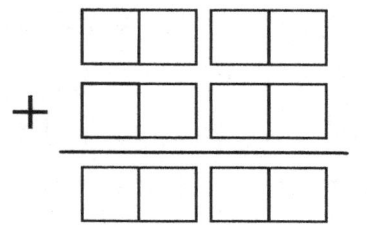

Problem 3. A $4 \times 4 \times 4$ cm^3 cube was cut into $1 \times 1 \times 1$ cm^3 cubes. Then, from all these cubes, a rectangular frame was made for a photograph 1 cube thick (similar to the one in the figure). It turned out that the area of the photo that fits into the frame is 216 cm^2. Determine the dimensions of the photo.

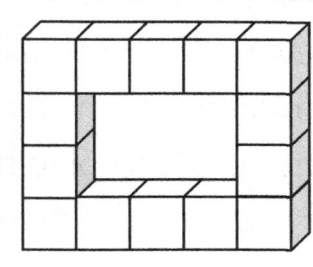

Problem 4. Cut the shape along the grid lines into 3 equal parts. (The cut parts may be rotated)

Problem 5. On Raduzhnaya Street, all the houses are located in a row. Each of the houses is painted in one of 5 different colors. It turned out that for two of these colors, it can be found neighboring houses painted in these two colors. What is the smallest number of houses on Raduzhnaya Street?

Problem 6. Some robots are trapped in a square-shaped trap made up of 36 square cells surrounded by columns (see figure). Each robot fires a laser simultaneously in 8 directions: the 4-sided and the diagonals. Robots are not in the line of fire of others. Likewise, they all fired at the same time and all the shots hit the columns. If the diagram shows the number of hits to each column. How many robots were there and where could they be?

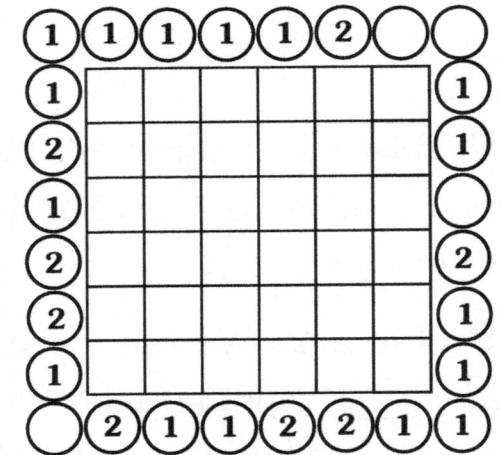

Problem 7. A mischievous monkey, a donkey, a goat and a black bear began to play in a quartet. They have violins (V), flutes (F), drums (D) and guitars (G). First, the bear and the monkey played guitars, the goat played the drums, and the donkey played the flute. Then they began to switch instruments. The bear every 15 minutes in this order: G-V-F-D-G-...., the monkey in the same order, but every 10 minutes. The donkey began to play in the order: F-G-V-D-F-... every 20 minutes. And the goat in the order D-F-V-G-D-... every 12 minutes. If they rehearsed for 2 hours and it is known that they only managed to play harmoniously when everyone had different instruments. How long did they manage to play harmoniously in 2 hours of rehearsal?

Problem 8. Five inhabitants of the island of gentlemen and liars were placed one after another. The last one (the fifth) said: "There are 4 liars in front of me". The fourth: "There are 3 liars in front of me". The third: "There are 2 liars in front of me". The second: "There is a liar in front of me". And the first was silent. How many of them are really liars? (If gentlemen always tell the truth and liars always lie)

ANSWER SHEET – EXAM 8

Problem 1	Answer
Problem 2	**Answer**
Problem 3	**Answer**
Problem 4	**Answer**

Problem 5	**Answer**

Problem 6	**Answer**

Problem 7	**Answer**

Problem 8	**Answer**

Exam 9

Problem 1. A boy removed the ornaments from a Christmas tree in a week. On Monday he removed 1 ornament and then each day he removed a number of ornaments equal to what he had removed on all the previous days together. If on Sunday he removed the last ornaments. How many ornaments were on the tree?

Problem 2. A boa (B), a monkey (M), an elephant (E), and a parrot (P) were weighed. The monkey wrote: Boa = 48 P, Elephant = 12 M, Monkey = 3 P, Boa = 4 M, Elephant = 36 P. It turned out that the monkey confused all the numbers, that is, the numbers were really the same, but they were all in different places (however, all the letters are written correctly). How many parrots do the boa, elephant, and monkey really weigh?

Problem 3. Put all the numbers from 0 to 9 in the cells to get the correct equality:

$$\boxed{} + \boxed{} \cdot \boxed{} - \boxed{} : \boxed{} = \boxed{}$$

Problem 4. A Christmas tree is drawn on graph paper. Cut it into 4 parts and form a square with them.

Problem 5. A group of children received 4 cards. Each card had one of the syllables PA, NA or MA written on it. It turned out that 13 children can form the word MAMA with their cards, 15 children can form the word PAPA and 17 children can form the word NANA. Likewise, 45 children can form the word PANAMA. How many children were there?

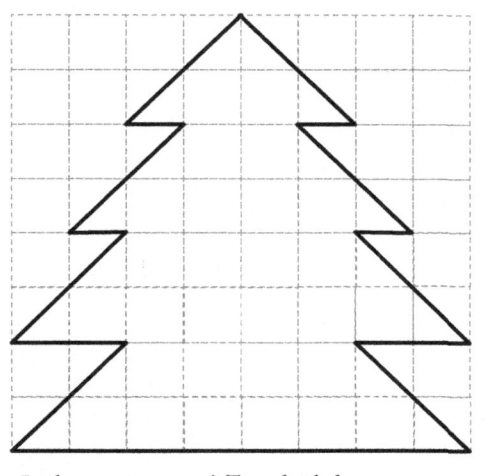

Problem 6. A rectangle was made from square cards (an example of a 6-card rectangle is shown in the figure). Then one side was reduced to its half and the other to its third part. If in the end 65 cards were removed. How many squares with a side of 4 cards can be formed from the original rectangle, without moving the cards? (in the figure it can be formed 2 squares with one side of 2 cards)

Problem 7. In the kingdom of the Full Moon, there are 9 cities as shown in the figure. The king wants to build direct roads between some cities so that they do not intersect outside the cities and there are exactly 4 roads leading out of each city. How could he do that?

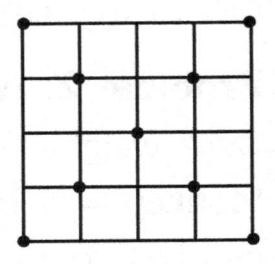

Problem 8. If a worm tells the truth, it turns green. And if it lies, it turns red. Once, two worms met. The first said: "We are both red". And then the second said: "If we had been silent, we would both be red now". Will the worms have the same color after this statement?

ANSWER SHEET – EXAM 9

Problem 1	Answer
Problem 2	**Answer**
Problem 3	**Answer**
Problem 4	**Answer**

Problem 5	Answer
Problem 6	**Answer**
Problem 7	**Answer**
Problem 8	**Answer**

Exam 10

Problem 1. Znayka multiplied two numbers and wrote down the resulting operation in encrypted form as: "MAKXIMUM". The signs for "multiplication", "equal", and each digit are represented by a letter. Likewise, different letters represent different signs or digits. What equality could Znayka have encrypted? Find at least one solution.

Problem 2. Vova swims a distance of 100 m in a 50 m long pool in 90 seconds, and in a 25 m long pool in 2 minutes. How long will it take for Vova to swim this distance in a 100 m long pool? Vova performs all the same actions and at the same speed.

Problem 3. If the smallest three-digit number that is not divisible by 4 is added the largest three-digit number divisible by 4. What is the result?

Problem 4. Kostya made a figure of three hexagons and wrote down numbers at all vertices, as in the left figure (he also wrote down a number at the central vertex; it is not known which one). Then Kostya increased the numbers at the vertices of one hexagon by a same number, then he increased the numbers at the vertices of the second hexagon also by a same number (possibly another), and then he did the same with the third hexagon. The right figure shows some of the numbers that were obtained. By how much has the number at the central vertex increased?

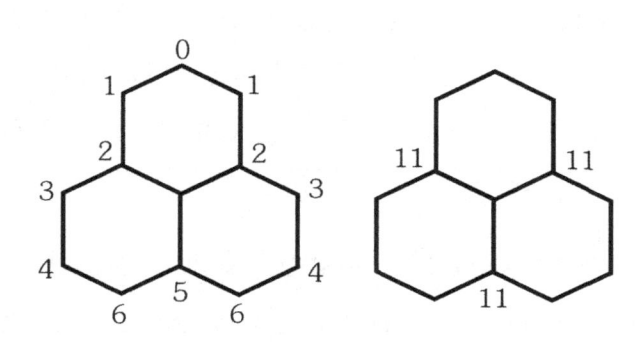

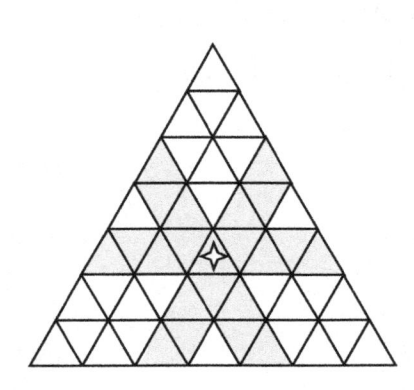

Problem 5. A park is divided into triangular sectors. If a flashlight is placed in one of the

253

sectors, it will illuminate in three directions, as in the figure. Light up the whole park by placing 3 flashlights.

Problem 6. Pinocchio, Pierrot and Artemon were playing with snowballs. Pinocchio threw 20 snowballs, Pierrot - 14 and Artemon - 8. It is known that all of Pierrot's snowballs flew past. Artemon threw a snowball only in response to a snowball that fell on him, and exactly half of Pinocchio's snowballs hit their target. How many snowballs hit Pierrot?

Problem 7. From home to school, Klim has three intersections with traffic lights. From the first to the second traffic light, Klim takes 2 minutes and from the second to the third traffic light also 2 minutes. Klim knows that at each traffic light, the yellow light is on for 1 minute, the green and the red for the same time. Likewise, he takes 1 minute, at the first traffic light, at the second - 2 minutes, at the third - 3 minutes. Once, Klim saw through the window that at 8:00 a.m. at all traffic lights the green light was turned on simultaneously. At what time must he be at the first traffic light to get to school without stopping at them? (Klim crosses the street in 5 seconds)

Problem 8. Sasha, Kolya, Masha and Olya live on different floors of a five-story building. Once Sasha said: "I live above all the others!", Masha: "And I am in the middle!", Kolya: "I live above Masha and below Olya". And Olya added: "Kolya told a lie. Between Kolya and me there is an apartment where neither of us lives". It turned out that those who lived on the odd floors had lied and those who lived on the even floors had told the truth. Where does each one live?

Problems – Level 4

ANSWER SHEET – EXAM 10

Problem 1	Answer
Problem 2	**Answer**
Problem 3	**Answer**
Problem 4	**Answer**

Mathematical Olympiads for Elementary School – *Problem Book*

Problem 5	**Answer**
Problem 6	**Answer**
Problem 7	**Answer**
Problem 8	**Answer**

Answers: Level 4

Answers - Level 4

Exam 1

1. 5 people.

2. 5 figures.

3. An example is shown as follows:

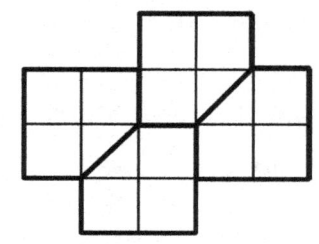

4. The heaviest is Hedgehog and the least heavy is Nyusha.

5. 30 minutes.

6. 10 rings.

7. (A) 11; (B) In the figure is shown an example:

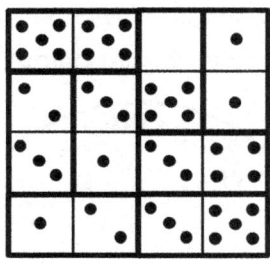

8. On the rocky island.

Exam 2

1. 2012 = 1717 + 295.

2. 5 years.

3. In the figure are shown two examples:

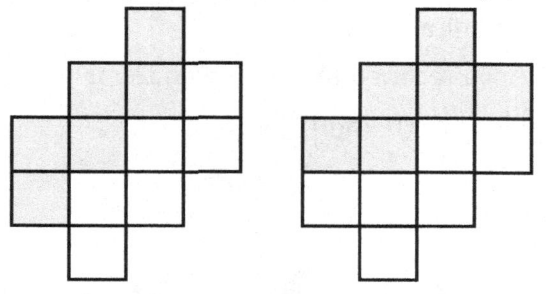

4. The filled table is shown below:

1	2	4	5	3	7	6
3	4	5	1	2	6	7
2	7	6	4	5	3	1
6	3	7	2	1	5	4
5	1	3	6	7	4	2
4	5	2	7	6	1	3
7	6	1	3	4	2	5

5. 88 *mm*.

6. 3 boxes.

7. The first grader solved three more problems than Borya.

Answers - Level 4

Exam 3

1. 1674 + 87 + 252 = 2013.
2. Tolya and Nikita.
3. Three times.
4. 51 minutes.
5. Anya received the piece with the number 2.
6. Here is an example:

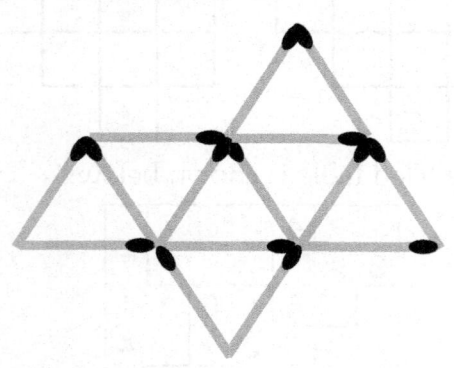

7. 20 minutes.
8. Gloria and Alex lied. Rico and Marty told the truth.

Exam 4

1. His name is Nikita Andreevich.
2. 33 km.
3. Some options are 2 × 4 cm or 2.5 × 6 cm rectangles. Any rectangle with a side of less than 3 cm will do.
4. An example is shown in the figure:

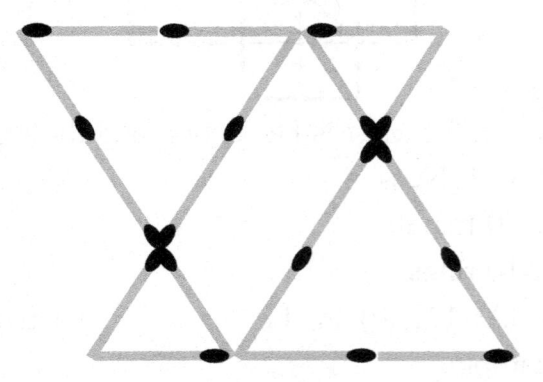

5. O+N+E+F+O+R+S+I+X = 35.
6. An example is shown below:

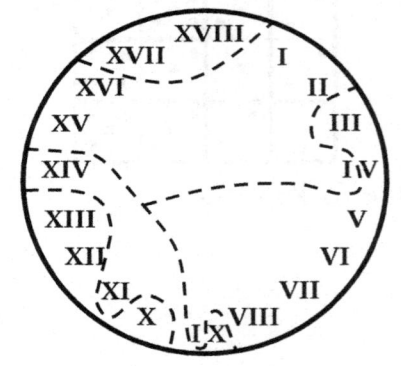

7. Popeye "the sailor" had spinach for breakfast.
8. Avoska was born on March 31.

Answers – Level 4

Exam 5

1. Masha - cookies, Sveta - apples, Vasya - chocolates, Petya - cake.
2. 36 solutions.
3. 5 chocolates.
4. In 1993 or in 2011.
5. An example is shown in the figure below:

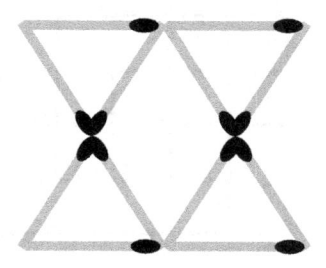

6. 12 ways.
7. He takes a handkerchief out of his pocket, puts the ball into his hat, and waves his wand.
8. There are 1 boy and 2 girls.

Exam 6

1. 2 coins.
2. An example is shown below:

4	2	3 >	1
1	3	4	2
3 >	1	2	4
2	4	1	3

3. See the example in the figure:

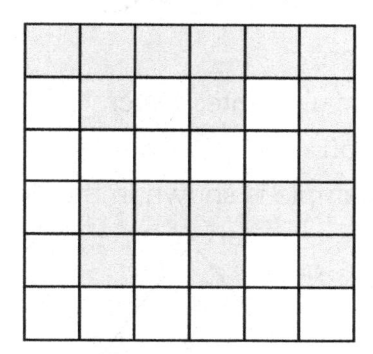

4. From top to bottom: F-W-Y-L-R-P-B-K-G-O.
5. $1305 \times 6 = 7830$.
6. There are 10 cookies in a packet.
7. The prince will defeat exactly 100 dragons and the wife will be lazy.

Answers – Level 4

Exam 7

1. For example, 132547698 or also 153428796 and there are other options.

2. An example is shown below:

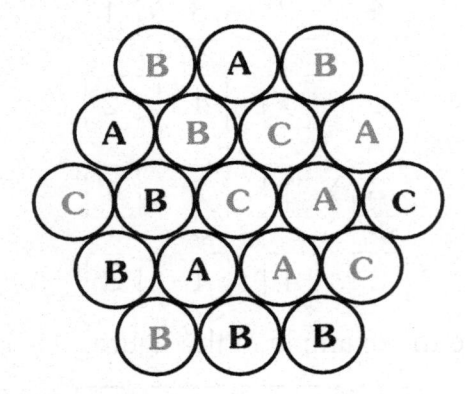

3. 3 stations.

4. Kamenyuk rotates faster.

5. 110 houses.

6. An example is shown in the figure:

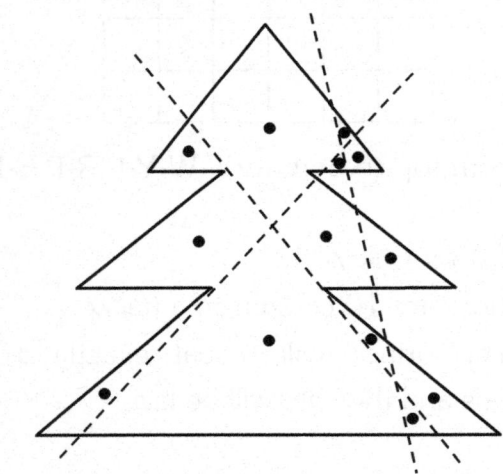

7. Valya broke the window.

8. Yes, it is possible.

Exam 8

1. More expensive are 3 chocolates.

2. A possible option is: 2560 + 3564 = 6124.

3. 12 cm × 18 cm.

4. An example is shown in the figure:

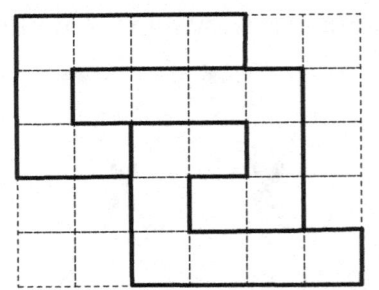

5. 11 houses.

6. 4 robots as shown in the figure below:

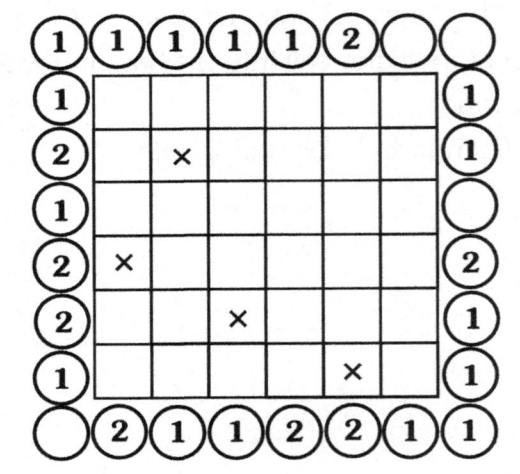

7. Every 8 minutes.

8. There are 4 liars.

Answers – Level 4

Exam 9

1. There were 64 ornaments on the tree.
2. Boa = 36 Parrots, Elephant = 48 Parrots, Monkey = 12 Parrots.
3. For example, 84 + 5 × 9 - 63: 7 = 120 or also 97 + 5 × 8 - 42: 6 = 130.
4. An example is shown in the figure:

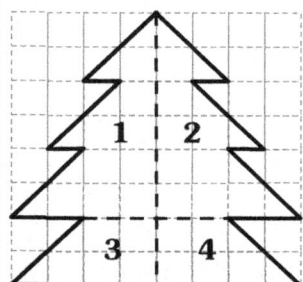

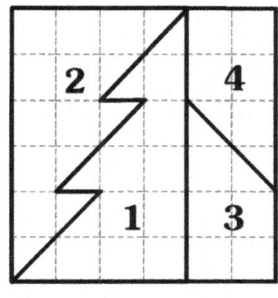

5. There were 45 children.
6. None.
7. An example is shown in the figure:

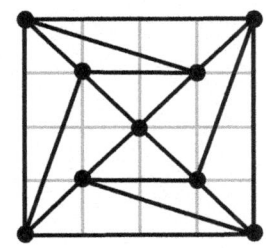

8. Yes, They will.

Exam 10

1. 28 × 9 = 252.
2. 75 seconds.
3. 996 + 101 = 1097.
4. It has increased by 12.
5. See the next figure:

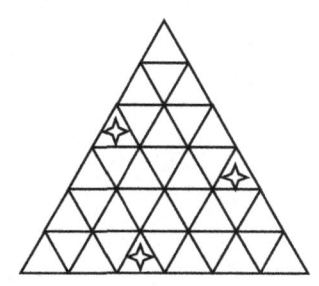

6. 2 snowballs.
7. At 12:00.
8. 1) Masha - 1°, Sasha - 3° and Olya - 5°, Kolya - 4° or also 2) Masha - 1°, Sasha - 4°, Olya - 3°, Kolya – 2°.

Answers - Level 4

Problems: Level 5

Problems – Level 5

Exam 1

Problem 1. Half a watermelon weighs 12 *kg* less than two of those watermelons. How much does a watermelon weigh?

Problem 2. Masha has 3 more candies than Petya and Petya has 5 less candies than Sveta. a) Which of the girls has more candy: Sveta or Masha? b) How many candies should one of them give to the other so that they have the same amount?

Problem 3. Pierrot planted several roses in a row along a straight line. Malvina planted two asters among the neighboring roses. A total of 19 flowers were planted. How many are roses?

Problem 4. Move a match to get the correct equality:

$$XI - IX = VII$$

Problem 5. Vasya wrote a number with three different digits. Olya wrote a number with two different digits. Tolya subtracted Olya's number from Vasya's number. What is the largest number he could get?

Problem 6. Dima folded a paper triangle along the dotted lines (see figure) and then cut the resulting small triangle as shown in the figure. How many pieces of paper did he get after doing this?

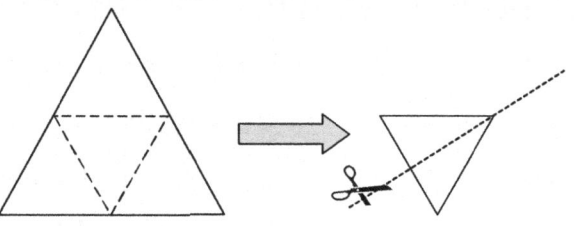

Problem 7. On a table there are three boxes, each with two different colored balls. In one there are blue and red balls, in another blue and yellow, and in the third yellow and red. Dunno made signs: "B-R", "B-Y" and "Y-R" which were placed on the boxes. But all the signs were placed

on the wrong boxes. Dunno drew a blue ball from the "B-R" box, a red one from the "B-Y" box, and a blue one from the "Y-R" box. What balls are in each box?

Problem 8. Vitalik made a pyramid with cubes. To prevent the pyramid from falling apart, he put a drop of glue between the adjacent faces. a) How many cubes are in the pyramid? b) How many drops of glue did he need?

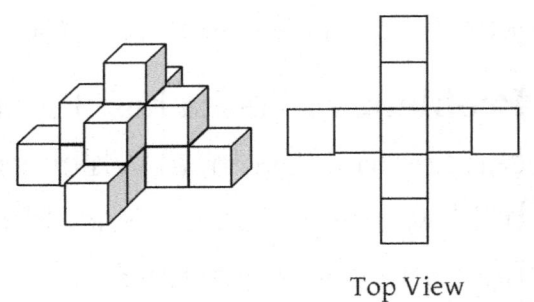

Top View

Problem 9. Between the digits 20112011 add a "+" and an "×" to get the largest possible result. You can use brackets if necessary.

Problem 10. 2010 identical dice are lined up. The faces in contact have the same number of dots. The beginning of the line is shown in the figure. How many dots can there be on the last face?

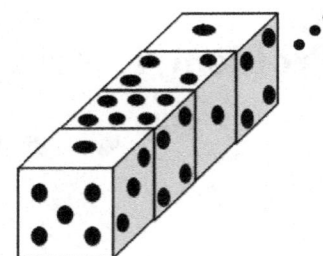

Problem 11. A barrel can be filled to the top by pouring 20 small, 5 medium and 8 large buckets of water, or 2 small, 2 medium and 11 large buckets of water. How many large buckets does it take to fill the barrel?*

Problem 12. Baba-Yaga gave Ivanushka the sword "kladenets" and said, "Watch out for the Three-headed Serpent <Gorynych>, it is cunning! Each head either shoots fire, or blows poisonous smoke, or turns them to ashes with a glance. The first head shoots fire, the second - blows smoke or shoots fire, and what the third does, I have forgotten". The old woman was kind, but very old: she did not remember everything correctly. What

---

*Editor's Note : The original problem was misstated. The necessary changes were made so that this interesting problem was not left out of our consideration.

Problems – Level 5

power of which head of the Gorynych Serpent will Ivanushka really have to face?

Problem 13. A strange shape with a square hole in the middle was cut from a square board (see the figure - the edges of the original board are marked with a dotted line). It turned out that the perimeter of the obtained figure is 8 cm larger than the perimeter of the original board. Calculate the area of the hole. (Note: all lines are parallel to the sides of the board)

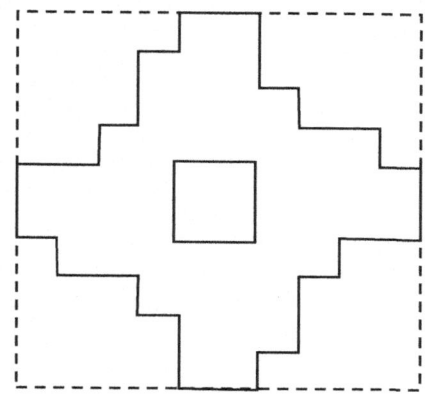

Problem 14. A pet fair featured cats of different breeds. It turns out that a tenth of the cats got a medal from some show, and among the Siamese cats, a seventh of them also got a medal. What was more at the fair: cats with medals or Siamese cats?

Problem 15. Dima has 12 coins and 5 pockets. Will he be able to arrange the coins in his pockets so that all the pockets have a different number of coins?

Problems – Level 5

ANSWER SHEET – EXAM 1

Problem 1	Answer
Problem 2	**Answer**
Problem 3	**Answer**
Problem 4	**Answer**

Problem 5	Answer
Problem 6	**Answer**
Problem 7	**Answer**
Problem 8	**Answer**

Problem 9	Answer
Problem 10	**Answer**
Problem 11	**Answer**
Problem 12	**Answer**

Problem 13	Answer

Problem 14	Answer

Problem 15	Answer

Problems – Level 5

Exam 2

Problem 1. Winnie the Pooh weighs 25 *kg* with five pots of honey and 19 *kg* with three pots of honey. How much does Winnie the Pooh weigh?

Problem 2. In the number 201220122012, cross out three digits to get the smallest possible nine-digit number.

Problem 3. Points A, B, C, D, E are marked in a straight line (in that order). It is known that AC = 29 *cm*, BD = 51 *cm*, BC = DE. What is the distance from A to E?

Problem 4. All the pages of Vitalik's book are numbered. The cover with the first pages came off. The initial page is number 7 and the last page is 26. How many sheets does the book now have?

Problem 5. Cut the 6 × 6 grid board along the lines of the grid, into four equal parts so that each part has exactly one square painted on it.

Problem 6. Masha multiplied a number by itself, obtaining a four-digit number. It is known that its digits of the hundreds and thousands are equal. Also, its ones and tens digits are also the same. What number did Masha multiply?

Problem 7. Gleb-Loba's great book of predictions says: "1) If it rains today, tomorrow it will be sunny. 2) If it snows today, it will rain tomorrow. 3) If it is cold today, tomorrow there will be wind. 4) If the sun rises today, tomorrow it will be hot. 5) If it is hot today, it will be cold tomorrow. 6) If it is cold today, tomorrow it will be cloudy. 7) If there is wind today, tomorrow it will snow. 8) If it is cloudy today, it will rain tomorrow". It turned out that in January all the predictions came true. On January 1 there was wind and sun. How was the weather on January 5?

Problem 8. There were 4 inhabitants of the Isle of Gentlemen and Liars in a room. They were asked, how many gentlemen are there among you? 4 different responses were received. How many gentlemen could there be in the room? List all the options. (Gentlemen always tell the truth and liars always lie)

Problem 9. Santa Claus bought magical fir and cedar seeds. Cedars grow 1.5 times taller than fir trees, but they grow for 9 hours. While fir trees grow for 2 hours. It is known that, Santa Claus planted cedar seeds at 12 noon and fir seeds at 2 pm. At what point were the trees the same height?

Problem 10. A city has nine districts. The length of the contour of each of them is 40 km. The districts are separated from each other by roads, the total length of which is 130 km. In addition, the city is surrounded by an outer ring road. What is the length of the latter?

Problem 11. The number 2012 is written on a whiteboard. In each operation, you can increase or decrease the number on the whiteboard by the product of two of its digits. Is it possible with this type of operation to obtain the number 2011 from 2012?

Problem 12. On the island of gentlemen and liars, there is a bus on a three-stop route. Three local residents, passengers on the bus, argued about a stop. The first: "Now we are at A. The next stop is B". The second: "No, we've already been at B. Now we are at C". The third: "If we are now at C. Only B will be missing". What is this stop?

Problem 13. Two rectangular sheets of paper are arranged as shown in the figure. It turned out that the length of the segment in bold is 10 cm. Compare and say which is larger: the area of the "overlapping" part (shaded part in the figure) or the "individual" part (four pieces remaining), if the length of a sheet of paper is 30 *cm*.

Problem 14. A group of children made a round. It turns out that five of them have two boys as neighbors, two more have a boy and a girl as neighbors, while the rest have two girls as neighbors. How many boys were there among the children?

Problem 15. A princess entered the castle through the main entrance and left through one of the seven auxiliary doors (which are marked on the plan with the letters A, B, C, D, E, F, G). The princess walked through 15 rooms without visiting any twice. What door could the princess go through to get out of the castle? List all the options.

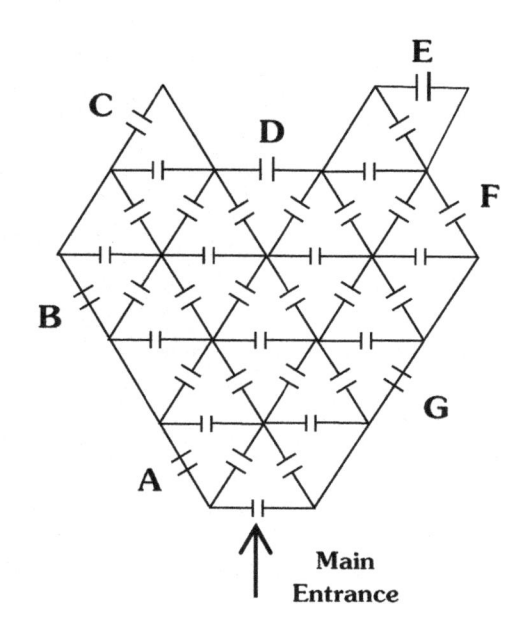

ANSWER SHEET – EXAM 2

Problem 1	Answer

Problem 2	Answer

Problem 3	Answer

Problem 4	Answer

Mathematical Olympiads for Elementary School – *Problem Book*

Problem 5	Answer

Problem 6	Answer

Problem 7	Answer

Problem 8	Answer

Problem 9	Answer
Problem 10	**Answer**
Problem 11	**Answer**
Problem 12	**Answer**

Mathematical Olympiads for Elementary School - *Problem Book*

Problem 13	Answer

Problem 14	Answer

Problem 15	Answer

Problems – Level 5

Exam 3

Problem 1. Which letter will be in third place from the end if all the letters in the word "MATEMATIKA" are written in alphabetical order?

Problem 2. Place the numbers 2, 3, 4, 6, 7, 8 at the vertices and midpoints of the sides of the triangle so that the sums of the numbers on each side equal 17.

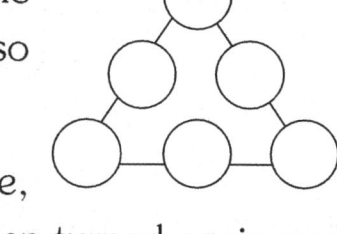

Problem 3. Vitalik wrote the 1 on the face of a cube, turned the cube and wrote the 2 on the next face, then turned again and wrote the 3, etc. Therefore, he numbered all the faces of the cube with numbers from 1 to 6. What is the maximum sum that the numbers of two opposite faces can have?

Problem 4. A rectangular surface of 4 *km* perimeter is divided into 4 rectangular garden sections, and a pool area in its center. What is the perimeter of the pool if the total length of the entrance doors (marked in bold in the figure) is 1700 *m*?

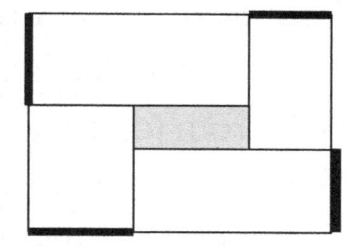

Problem 5. In a card board game, before starting, a card is placed on the table and the rest is distributed equally among the players. What is the smallest number of cards in this game so that two, three, four, five or six people can play it?

Problem 6. Three swans, three river turtles, and four fish came together to tow a small boat. It turned out that for this they need to organize themselves in a circle so that there are not three swans, not three river turtles, and not three fish together. Also, it is recommended that three consecutive places are not occupied by a swan,

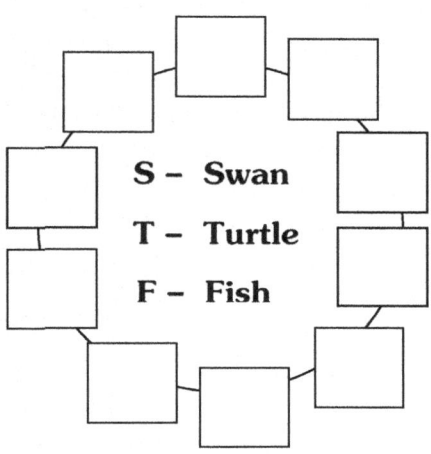

S – Swan
T – Turtle
F – Fish

a river turtle and a fish in any order. Distribute them in a circle so that all these conditions are met.

Problem 7. Find a number such that if you multiply one third of it by its fifth, you get that number.

Problem 8. Piglet, Eeyore, and Winnie the Pooh were counting carrots on two of the Rabbit's beds. When they finished counting, they said the following – Piglet: "On the first bed there are more than 18. On the second bed, no more than 14". Eeyore: "On the first bed there are less than 20. On the second bed there are 14". Winnie the Pooh: "On the first bed there are 17. On the second bed there are more than 14". It is known that one of them was wrong when counting both times, and the other two were right. How many carrots were on the beds?

Problem 9. Lamps are installed on both sides of the metro's escalators, at regular intervals. All lamps are numbered starting from 1. On the left side, from top to bottom, and on the right side, from bottom to top. While Dima was on the escalator, he saw on one side a lamp with the number 7, and on the other one with the number 17. How many lamps are there?

Problem 10. On the petals of a seven-colored flower it is written which petal should be plucked next. On the red one: "light blue", on the orange one: "green", on the yellow one: "violet", on the green one: "blue", on the light blue one: "yellow", on the blue one: "orange", on the violet one: "green or red". Only one petal can be plucked at a time. Zhenya managed to tear off all the petals. Which petal did she pluck first?

Problem 11. Kostya has four large and four small triangles (see figure). Help Kostya to build a square without holes, using all these triangles without overlapping.

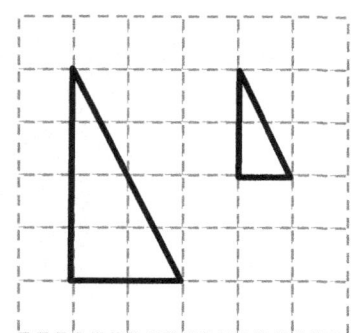

Problem 12. Three bearded sages argued about who had the longest beard. Each of them said – the first: "I have the longest beard among us!", the second: "No, mine is longer than yours!", the third: "At least one of you is wrong". Which of the sages has the shortest beard, if the length of all the beards is different and only the sage with the longest beard tells the truth?

Problem 13. A sports team consists of six people. All of them participated in five competitions, competing together. Could it be that the sum of the places occupied by each is the same?

Problem 14. Aunt Grusha sells zucchini. She sells three zucchini for $ 5, four zucchini for $ 6, and five zucchini for $ 7. Aunt Grusha does not sell zucchini in any other quantity. If she sold 100 zucchini for $ 160 yesterday. How many sales did Aunt Grusha make yesterday?.

Problem 15. There are 20 Olympiad participants in a room. In addition, it is known that among 10 of them there are 3 classmates. Is it true that there are necessarily 5 people of the same class in the room?

ANSWER SHEET – EXAM 3

Problem 1	Answer
Problem 2	**Answer**
Problem 3	**Answer**
Problem 4	**Answer**

Mathematical Olympiads for Elementary School – *Problem Book*

Problem 5	Answer
Problem 6	**Answer**
Problem 7	**Answer**
Problem 8	**Answer**

Problem 9	Answer

Problem 10	Answer

Problem 11	Answer

Problem 12	Answer

Problem 13	Answer

Problem 14	Answer

Problem 15	Answer

Problems – Level 5

Exam 4

Problem 1. Tolya thought of a number. First, he added 1 to it, then multiplied the result by 2, and finally subtracted 5, getting 17. What number did he think of?

Problem 2. If Petya gives half of his candies to Masha, then Masha will have 5 more candies than Petya. How many candies does Masha have?

Problem 3. Place the numbers 2, 6, 0, 1, 2, 0, 1, 4 at the vertices and midpoints of the sides of the square so that the sums of the numbers on each side are divisible by 3.

Problem 4. Lumberjack Petrovich cuts four 5-meter logs into meter-long logs in 20 minutes, and Lumberjack Palych during this time cuts seven 3-meter logs into meter-long logs. Which one of them will cut a 10-meter log into smaller logs faster?

Problem 5. A box contains balls of various colors (at least 7). If five balls are drawn, then among them there will definitely be two balls of the same color. And if seven balls are drawn, there will definitely be two balls of different colors. What is the maximum number of balls in the box? And what is the minimum?

Problem 6. Fedya takes in the same number of vitamins every day. Vitamins are sold in large, medium, or small jars. A large jar contains three times more than a small one, and a medium one contains twice as much as a small one. Fedya finishes a large jar completely in exactly 50 days. A small jar only lasts 16 days, with some vitamins left. How many days will a medium jar last?

Problem 7. In the figure, the gears are coupled to each other by their teeth and belts. Indicate in which direction the last gear will rotate.

Problem 8. Place two signs of arithmetic operations (+, −, × or ÷, they can be repeated) between the digits shown below, so that the value of the expression is the largest possible. The numbers formed cannot start with 0.

$$2\ 6\ 0\ 1\ 2\ 0\ 1\ 4$$

Problem 9. There are three people in a meeting, each of them either always tells the truth or always lies. The first said: "There is a liar among us", the second said: "There are two liars among us", and the third said: "There are three liars among us". Who is who?

Problem 10. Cut the shape along the grid lines into four equal parts so that each part contains exactly one painted cell.

Problem 11. Petya, Vanya, Lena, Sasha and Misha live on Lipovaya Street. Petya and Vanya live on opposite sides of the street, Misha and Sasha live on the same side of the street, Sasha and Petya also live on opposite sides of the street, Lena and Misha also live on opposite sides of the street. Do Vanya and Lena live on the same side of the street or on opposite sides of it?

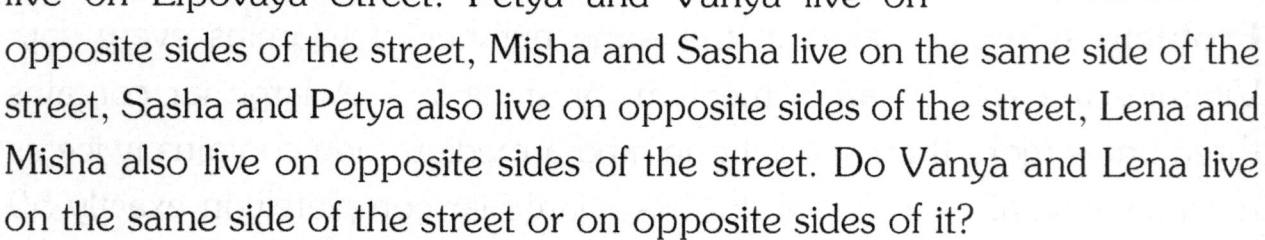

Problem 12. Dunno solved an example of multiplication of two two-digit numbers on the board. Also, the four digits of the factors are different. Then Dunno erased these digits and replaced them with letters,

but did not touch the multiplication sign, the equal sign, and the result, obtaining AB × CD = 9000. Show that Dunno would have been wrong somewhere.

Problem 13. Kostya has six coins, identical in appearance, of which 4 are real and weigh the same, and two are false: one is lighter and the other is heavier. If you put the fake coins together, they weigh the same as two real ones. How can Kostya find two real coins in two weighings with a two-plate scale?

Problem 14. Last week, Kolya, Tolya, and Olya bought 5 identical packets of sweets. They opened all the packets, stacking the sweets in a pile and began taking one at a time. It turned out that all three got the same amount. This week they bought the same sweets, but now 13 packets, and again they piled them in a single pile. Kolya said: "I wonder if this time we will be able to divide the sweets equally". Olya replied, "Maybe it will work, but maybe not. It depends on the number of sweets in the packet". Tolya objected, "No, it doesn't matter how many treats are in the package! Once it worked with 5 packages, it will work with 13 packages!". Who is right: Olya or Tolya?

Problem 15. Five oaks grow along a straight alley (the distance between the oaks is not necessarily the same), the distance between the first and the last oak is 28 meters. In the middle between the first and second oak, Rabbit planted a carrot. In the middle between the second and third oak trees, Winnie the Pooh planted a rose. In the middle between the third and fourth oak, Piglet buried an acorn. In the middle between the fourth and fifth oak, Eeyore planted a thistle. Christopher Robin measured the distance between the carrot and the thistle as 20 meters. What is the distance between the rose and the acorn?

Problems – Level 5

ANSWER SHEET – EXAM 4

Problem 1	Answer
Problem 2	**Answer**
Problem 3	**Answer**
Problem 4	**Answer**

Mathematical Olympiads for Elementary School – Problem Book

Problem 5	Answer
Problem 6	**Answer**
Problem 7	**Answer**
Problem 8	**Answer**

Problems – Level 5

Problem 9	Answer

Problem 10	Answer

Problem 11	Answer

Problem 12	Answer

Mathematical Olympiads for Elementary School – *Problem Book*

Problem 13	**Answer**

Problem 14	**Answer**

Problem 15	**Answer**

Exam 5

Problem 1. Write the smallest even number of six different digits.

Problem 2. Some biologists planted trees: pine, spruce, and fir. As the botanist Fernnikov calculated, out of 5 planted trees there is at least one pine, out of 6 planted trees there is at least one spruce, and out of 8 planted trees there is at least one fir. How many trees of each species have biologists planted?

Problem 3. The perimeter of a square is 4 times the perimeter of another square. How many times the area of the smaller square is the larger square?

Problem 4. Each of the children of the same family declared that they have the same number of brothers and sisters. It is known that 5 people were wrong. a) How many brothers could there be in the family? b) How many children could there be in the family?

Problem 5. Little Vova noticed that in January he only thought about gifts, summer and the next Olympiads. Likewise, he thought about the gifts all January except the last 7 days; in the summer, all the month but from January 8; and in the Olympiads, only on the days of the month, in which there is a two. How many difficult days did Vova have in January, in which he thought about everything at once?

Problem 6. All the corners of a wooden cube were cut, as shown in the figure. a) How many edges does the resulting shape have? b) How many vertices does it have? c) And how many faces does it have?

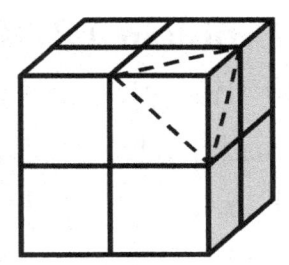

Problem 7. Some children are dancing in a round. It turned out that each boy has a boy on one side and a girl on the other side. And each girl has boys on both sides. How many girls are there, if there are 12 boys?

Problem 8. A Contest of Wisdom was held in the City of Flowers. When asked who won, four men answered like this:
The first: Zemaforos from the Solar City or Kusachkin from Zmeyovka.
The second: Gaikin from Prostokvashino or Zemaforos from Zmeyovka.
The third: Zemaforos from Prostokvashino or Gaikin from Zmeyovka.
The fourth: Kusachkin from Prostokvashino or Prostofilin from the Lunar City or Zemaforos from Tsvetochny. Determine who won and what city he is from, if each answered the correct name of the winner or the correct city of the winner, but not both.

Problem 9. If a three-digit number is subtracted by the sum of two of its digits, we get 777. Find the original number.

Problem 10. Paint the squares on an 8×8 grid board with 5 colors so that any 1×5 strip contains all five colors, and any 2×2 square and any 1×4 strip contains four different colors.

Problem 11. A seller has envelopes in packs of 100 pieces. He can count 5 envelopes per second. If he needs to prepare three packs with 50, 70 and 80 pieces, respectively. What is the minimum time in which the seller can prepare the packs?

Problem 12. Karabas – Barabas has 1 gold, 2 silver and 2 bronze coins. If the coins are real, then 1 gold coin weighs the same as 2 silver coins, and 1 silver coin weighs the same as 2 bronze coins. How can Karabas – Barabas find 1 counterfeit coin among these 5 coins, which differ in weight from the real one (but it is not known whether it is lighter or heavier), with only two weighings on a two-plate scale?

Problems – Level 5

Problem 13. A sports complex consists of 6 identical courts located as shown in the figure. The director of the complex ordered to enclose the entire complex around the perimeter with a fence. What is the length of the fence if the perimeter of a court is 170 m?

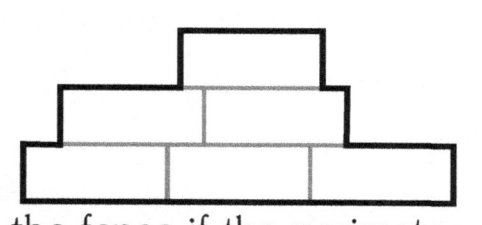

Problem 14. Sasha bought nine office supplies for a total of 3 rubles. It is known that any two consecutive purchases cost the same in total. Show that the most expensive purchase cannot cost more than 70 kopecks (1 ruble = 100 kopecks).

Problem 15. Donkey Eeyore received a mathematical cube on his birthday, on each face is written "2", "2 × 2" or "2 × 2 × 2". Eeyore threw the die on the table and calculated the sum of the values on all the visible faces, obtaining 16. When the Owl did the same, he obtained 20. On how many faces of the cube is written "2 × 2"?

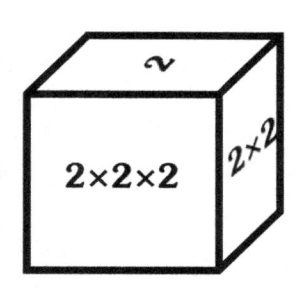

ANSWER SHEET – EXAM 5

Problem 1	Answer

Problem 2	Answer

Problem 3	Answer

Problem 4	Answer

Mathematical Olympiads for Elementary School – Problem Book

Problem 5	Answer

Problem 6	Answer

Problem 7	Answer

Problem 8	Answer

Problem 9	Answer
	|//
Problem 10	Answer
Problem 11	Answer
Problem 12	Answer

Problem 13	**Answer**
Problem 14	**Answer**
Problem 15	**Answer**

Problems – Level 5

Exam 6

Problem 1. Write the smallest five-digit number that is divisible by 3.

Problem 2. Place the signs of arithmetic operations (+, −, × or ÷, they can be repeated) and an equal sign between the digits shown, to obtain the correct equality.

3 1 0 1 2 0 1 6

Problem 3. A square sheet of paper has a folded corner such that the vertex coincides with the center of the square, as shown in the figure. The area of the resulting pentagon is 2 cm^2 less than the area of the original square. What is the side of the square?

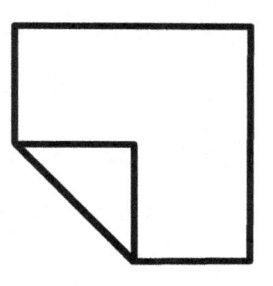

Problem 4. Divide a nonagon or regular eneagon into exactly 7 isosceles triangles.

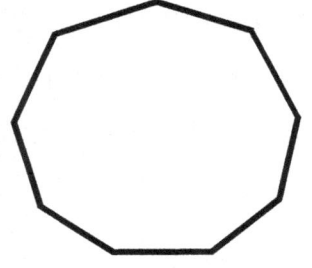

Problem 5. In the fifth grade classroom, 7 students eat ice cream every day, 9 students eat ice cream every other day, and the rest do not eat ice cream at all. Yesterday, 13 students in this classroom ate ice cream. How many students will eat ice cream today?

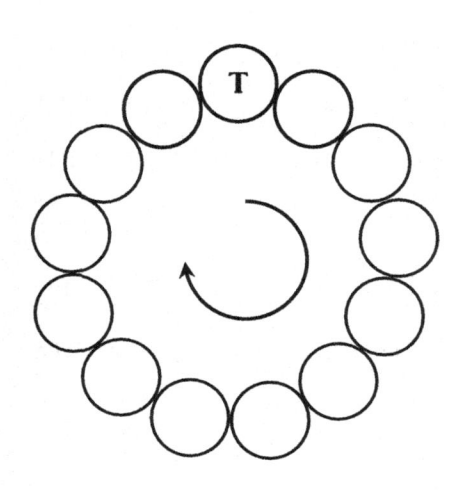

Problem 6. Usually Ilya Yakovlevich rests only on Saturdays and Sundays. But in February 2010, Ilya Yakovlevich received a leave of absence for 12 days in a row. What is the maximum and minimum number of uninterrupted rest days in February? (If Saturday or Sunday falls on leave days, they will be considered leave days.)

Problem 7. Ocean, Terry, and their eleven friends formed a circle to choose who was going to play chess. Terry counts, starting with the neighbor on the left and continuing in a clockwise direction. Where does Ocean have to be for everyone to get elected except him and Terry? If it is known that Terry always uses the following counting rule: "5 – 5 – 2 – 6 – 5 – 5 – 2 – 7"

Problem 8. From 4 solid shapes, where each one consists of 4 cubes, a rectangular parallelepiped was built as shown in the figure. If each shape has a different color. What does the white shape look like?

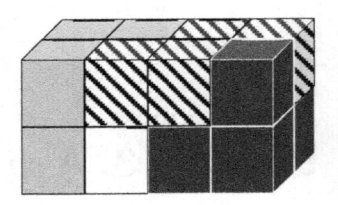

Problem 9. Vanya walked down the left side of a street and considered adding the digits of the house numbers she saw on this side. At some point (when she passed by at least two houses) she got 51. At that moment she stopped and across the street she saw number 17. With what maximum number of houses could she do the addition? And what is the minimum? (Consider that the houses are numbered in an orderly fashion, with the even numbers on one side of the street and the odd ones on the other side).

Problem 10. The boys Pasha, Kolya, Tolya and the girls Masha and Sveta agreed to meet at a skating rink. But in the end, not all of them went. When asked: "Who went there?", The children answered like this: Tolya: "we were four". Masha: "there were more boys than girls". Pasha: we were three. Sveta: "we were both with Masha". Kolya: "Tolya was not there". Who went to the skating rink if only those who were told the truth and the rest lied? List all possible options.

Problem 11. Will it be possible to place the numbers 3, 1, 0, 1, 2, 0, 1, 6 at the vertices of a cube so that the sums of the numbers on all its edges are different?

Problem 12. A pentagon was drawn on graph paper and partially painted gray (see figure). Which part of the pentagon has a larger area: the painted one or the unpainted one?

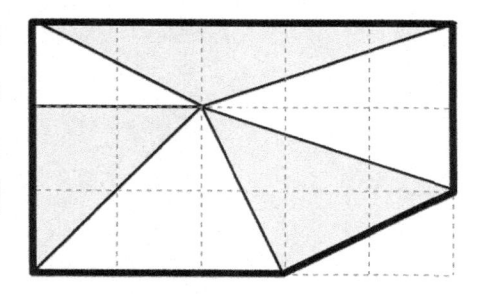

Problem 13. Koschey the Immortal decided to fill a chest with emeralds. On the first day he put 1 emerald in the empty chest. The next day, he put 2 emeralds there, and so on; each day he put 1 more emerald in the chest than the day before. However, on the second night, Baba-Yaga stole 1 emerald from the chest and each subsequent night she stole 1 more emerald than the night before. As soon as there are 2016 emeralds in the chest, Koschey will seal and hide it, and Baba-Yaga will not be able to steal any more. What day will it happen?

Problem 14. Carlson received a box of chocolates for his birthday. Also, Krister ate fewer chocolates and Gunilla more. Sasha ate an even number of chocolates, 3 times more than Krister and 2 times less than Gunilla. Carlson ate all the other chocolates. Could there be 65 chocolates in the box?

Problem 15. Pasha, Kolya, Lesha and Sasha played in a chess tournament in one round (all against all). It turned out that everyone obtained a different number of points and there were no ties. After the tournament, everyone talked about their participation – Pasha: I made the most points. Kolya: Lesha made more points than Pasha. Lesha: Kolya and Pasha have the same points together as Sasha. Sasha: My score is better than Lesha's. Determine the place occupied by each of them in the tournament, if it is known that they all lied. (For a victory in chess, 1 point is awarded, and for a defeat 0 points)

ANSWER SHEET – EXAM 6

Problem 1	Answer
Problem 2	**Answer**
Problem 3	**Answer**
Problem 4	**Answer**

Mathematical Olympiads for Elementary School – *Problem Book*

Problem 5	Answer
Problem 6	**Answer**
Problem 7	**Answer**
Problem 8	**Answer**

Problem 9	Answer
Problem 10	**Answer**
Problem 11	**Answer**
Problem 12	**Answer**

Problem 13	Answer
Problem 14	Answer
Problem 15	Answer

Exam 7

Problem 1. At what digit does the product of all numbers divisible by 2017 and less than 20170 end?

Problem 2. Yegor and Artyom have 45 stamps together. Half of Yegor's stamps are equal to one-third of Artyom's stamps. How many stamps does each child have?

Problem 3. Three identical squares were joined to each other by their sides (without overlapping) so that a rectangle resulted. What is the area of the rectangle if its perimeter is 48 *cm*?

Problem 4. In a group of children it is observed that among any four of them is Ivan. And among any three of them there is a girl. What is the largest number of boys in this group?

Problem 5. Petya added three consecutive numbers and got a number with different digits. While he was copying the result into a notebook, she forgot to write down the last digit and wrote down 1046. What are these three numbers?

Problem 6. If one more cell is added to the figure of four cells with the shape of the letter L, a figure with an axis of symmetry can be obtained. How many ways can this be done?

Problem 7. An electronic clock shows the time in 24-hour format. What is the maximum number of minutes in a row in which four digits will be displayed on the screen in order a) increasing; b) not decreasing? (Instead of 24:00, the clock shows 00:00)

Problem 8. On a paper pyramid, a point was chosen on each face, and a dotted line was connected to the vertices of the faces. Likewise, the pyramid was cut out along all the dotted lines. How many pieces of paper were obtained?

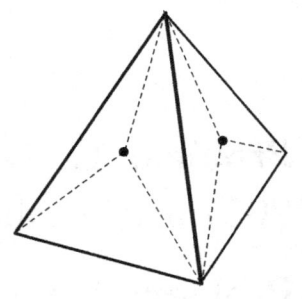

Problem 9. Olya, Vasya, Masha, and Petya are fourth, fifth, sixth, and seventh grade students. When asked who is older than who, the kids said the following; Olya: "Masha is older than Petya", Vasya: "Olya is younger than Petya", Masha: "Petya is older than Vasya", Petya: "Masha is younger than Olya". Later it was learned that if someone spoke of a student older than them, they was lying. If all other statements were correct, determine to what grade each belongs.

Problem 10. Lyosha wrote on the whiteboard a natural number less than 1000. For each operation, divide the current number on the board by 2, if it is even, and write the result of the division in place of the previous number. If the number on the whiteboard is odd, add 1 to it and also write the result of the addition in place of the previous number. What is the maximum number of operations after which Lyosha can get the number 1 on the whiteboard for the first time? (For example, if 5 was written, we get 1 after 5 operations: 5→6→3→4→2→1)

Problem 11. Ostap has 4 siblings. Once his mother brought 50 sweets and put them on a plate. Ostap took a number of sweets. And then the other siblings took of the remaining sweets. Each sibling took at least 2 times more than the previous one. What is the most candy that Ostap can have?

Problem 12. Grisha wrote the number 16 on a whiteboard, and in each operation, he adds the greatest prime divisor to the number on the whiteboard (erasing the old number and writing the new one). From the number 16, he will get the sequence: 16-18-21-28-35 -... Could a number like 1000 ... 000 appear on the whiteboard at some point?

Problem 13. There are three houses along the shore of a circular lake and they belong to Owl, Rabbit, and Winnie the Pooh. Piglet planted an oak tree exactly in the middle between the houses of the Owl and the Rabbit. And exactly in the middle between the houses of Winnie the Pooh and the Rabbit, bees live in a hive. Now Piglet is exactly in the middle between Owl and Winnie the Pooh houses. If he goes to the oak, visiting Owl, he will walk 17 *km*, and if he goes to the oak, visiting Winnie the Pooh and Rabbit, he will walk 35 *km*. What is the distance along the shore from Winnie the Pooh's house to the bee hive?

Problem 14. On January 31, 2016, a crowd of men approached Dr. Pilyulkin complaining of poor health. Pilyulkin ordered everyone to take vitamins during February: 1 tablet, once a day. Dunno, as always, was late and didn't go to the doctor until February. Pilyulkin prescribed the vitamins from the next day. It turned out that in February the men (including Dunno) took 2017 tablets. What day in February did Dunno visit Dr. Pilyulkin?

Problem 15. 2017 inhabitants of the island of gentlemen and liars stood in a circle (gentlemen always tell the truth and liars always lie). Each of them was asked to name their neighbor on the right, and each answered "gentleman" or "liar". Could exactly 2000 answers be "gentleman"?

ANSWER SHEET – EXAM 7

Problem 1	Answer
Problem 2	**Answer**
Problem 3	**Answer**
Problem 4	**Answer**

Problem 5	Answer
Problem 6	Answer
Problem 7	Answer
Problem 8	Answer

Problem 9	Answer

Problem 10	Answer

Problem 11	Answer

Problem 12	Answer

Problem 13	Answer
Problem 14	Answer
Problem 15	Answer

Problems – Level 5

Exam 8

Problem 1. Place arithmetic signs (+, −, × or ÷, they can be repeated) and brackets between some digits to get the correct equality:

$$2\ 0\ 1\ 7\ 7\ 1\ 0\ 2\ 2\ 0\ 1\ 7 = 2018$$

Problem 2. A car entered a tunnel at a speed of 5 *m/min* in the middle of a traffic congestion. First, he drove for 11 *min*, then stopped for 2 *min*, then he drove for 10 *min*, and stopped for 3 *min*, then he drove for 9 *min*, and stopped for 4 *min*, and so on, until the tunnel ended. He reached the end of the tunnel in 55 *min*. How long is the tunnel?

Problem 3. Yegor wrote several consecutive two-digit numbers. Resulting that each digit from 0 to 9 was written at least once. What is the least number of numbers Yegor could write?

Problem 4. A Rabbit's clock has no numbers and its hands are the same length. One day the watch fell off and the Rabbit hung it up again without paying attention to how he hung it. What time does the clock show?

Problem 5. Kostya places the dominoes in a chain according to the rules of dominoes, choosing each time among the remaining dominoes, the one with the maximum sum of points. A) How many dominoes will the chain have? B) What will be the last domino in the chain?

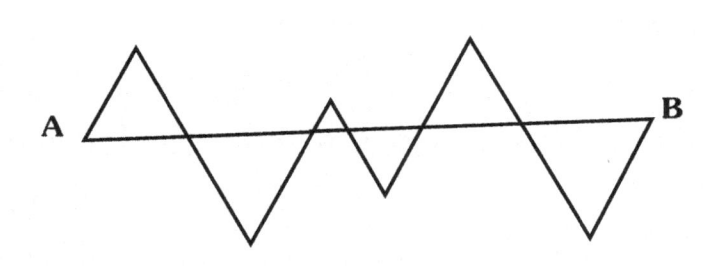

Problem 6. A 7-piece polyline with ends at points A and B is crossed by line segment AB. Obtaining that all the resulting triangles are equilateral. The

length of segment AB is 12 *in*. Calculate the length of the polyline. (Equilateral triangle: a triangle in which all the sides are equal)

Problem 7. There is a chip in one of the cells of the checkerboard. It can only be moved to an adjacent cell. If the chip went through all the cells on the board once and stopped. In which cell on the board could it have stopped? Mark all possible cells.

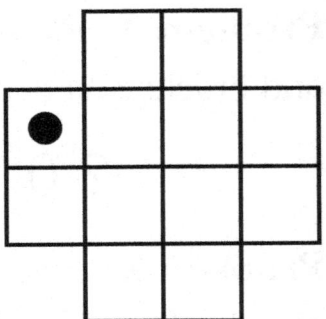

Problem 8. A Martian zoo houses dragons with three heads and four legs and chuchundras with five legs and six heads. Alice counted that there were 126 heads and 123 legs in all. How many dragons and how many chuchundras live in the Martian zoo?

Problem 9. There are two wire grids with 4 square cells. Cut the grids into 4 identical parts (each into 2 parts) and construct a 9-cell wire grid, as shown in the figure.

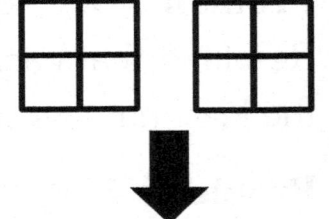

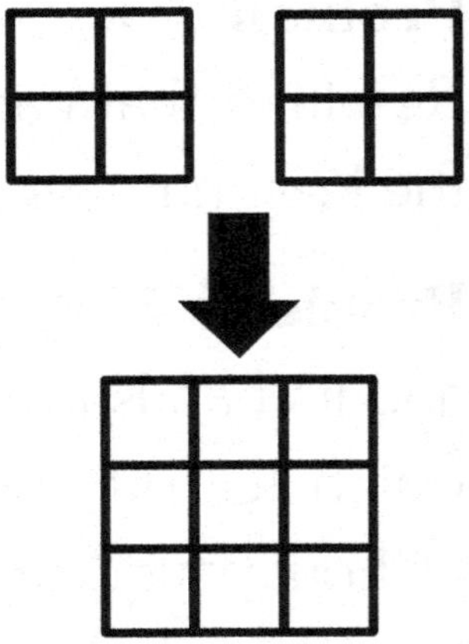

Problem 10. There were 11 people in a room, inhabitants of the island of gentlemen and liars (gentlemen always tell the truth and liars always lie). The first one said: "There is a liar among us". The second one: "There is a gentleman among us". The third one: "There are 2 liars among us". The fourth one: "There are 2 gentlemen among us". And so on, the tenth one: "There are 5 gentlemen among us". And the last one said nothing, how many liars could there really be in the room?

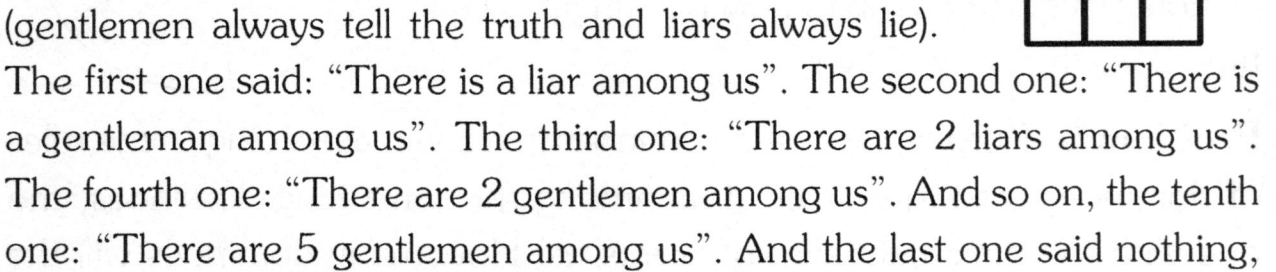

Problem 11. Pinocchio has 5 coins that look exactly the same. Of these, 2 are fake, and both are lighter than the real ones and weigh the same. Pinocchio must pay a real coin for lunch. The restaurant owner allows you to use a two-plate scale exactly once. How can Pinocchio find the real coin? (he can put any number of coins on the scale)

Problem 12. A rectangle was drawn on graph paper, the sides of which follow the lines of the grid. The rectangle was cut into four rectangles with two straight cuts, also along the lines of the grid. Petya, a fifth grader, calculated that three of these rectangles have areas of 4 cm^2, 8 cm^2, and 16 cm^2. What is the area of the original rectangle? Find all the possible options and show that there are no others.

Problem 13. For the preparation of a potion to turn a pink pony into a blue unicorn, milk from one bird should be boiled over low heat for 33 minutes, then immediately add stardust and cook for another 6 minutes, and finally add drops of dew from poppies and five clever thoughts. Harry Potter has two hourglasses, 4 *min* and 7 *min* respectively. Help him use these clocks to measure the required intervals and brew the potion.

Problem 14. Three thieves found a treasure of 9 emeralds of: 3, 4, 7, 8, 9, 12, 17, 21, 22 grams. They want to divide the emeralds so that everyone gets three pieces, such that the one received by the oldest weighs 2 times more than the one in the middle and the one in the middle weighs 2 times more than the one received by the youngest. Can they divide the emeralds in this way? If so, how? If not, why not?

Problem 15. In the city of Octopus, a subway with 8 stations on it was built. Likewise, 4, 3, 3, 3, 2, 2, 2, 1 metro lines depart from them, respectively (one line connects exactly two stations). A line for repairs was closed. Could it be that the metro map now consists of two identical independent sections?

ANSWER SHEET – EXAM 8

Problem 1	Answer

Problem 2	Answer

Problem 3	Answer

Problem 4	Answer

Problem 5	**Answer**

Problem 6	**Answer**

Problem 7	**Answer**

Problem 8	**Answer**

Problem 9	Answer

Problem 10	Answer

Problem 11	Answer

Problem 12	Answer

Problem 13	Answer

Problem 14	Answer

Problem 15	Answer

Problems – Level 5

Exam 9

Problem 1. Put the numbers 2, 7, 1, 2, 0, 1, 9 in the cells so that the product is as large as possible:

☐☐☐☐ × ☐☐☐

Problem 2. Rostik is training to play the drums: every 10 seconds he hits the kick drum, every 4 seconds, the cymbal, and every 7 seconds, the drum. He started hitting the kick drum and cymbal, then after 3 seconds he added the drum. If Rostik played for 4 minutes. How many times during this time did he hit the kick drum, the cymbal, and the drum at the same time?

Problem 3. Nikita replaced in a correct example of addition, the same digits for the same letters, and different digits for different letters obtaining the following puzzle: M + A + T + E + M + A + T + I + K + A = EE ¿What is the largest digit that can be replaced by the letter E?

Problem 4. Jan wrote all the consecutive numbers from 1 to 50 with no blanks. Then he found the five most common digits in this sequence and crossed out all of those digits. Which digit of the new sequence is A) the first one? B) the last one?

Problem 5. Draw 6 points and connect with segments so that at each point 4 segments intersect and there are no other points of intersection.

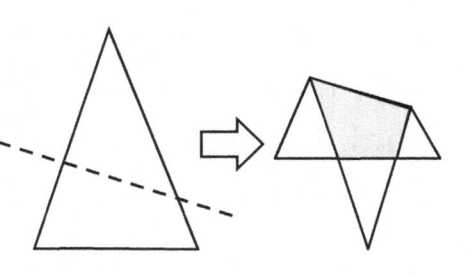

Problem 6. On a paper triangle, a line was drawn dividing the area of the triangle in half, and then the triangle was folded along this line. It turned out that the area of the "overlapping part" (gray area in the figure) is

equal to the area of the "individual parts" and is 12 cm^2 less than the area of the original triangle. Find the area of the lower small triangle.

Problem 7. When it is noon in Belgrade, it is 11 p.m. in Kamchatka. At the same time in Boston it is 6 a.m., and in Los Angeles it is 3 a.m. the same day. On January 10 at 8 p.m. Misha emailed a photo from Boston to Belgrade to Vova (photos are emailed almost instantly). After 14 hours, Vova emailed it to Rodion in Kamchatka. The next morning at 8 o'clock, Rodion sent it to Grisha to Los Angeles, also by email. At what time and on what date was the photo received by Grisha?

Problem 8. The brim of a witch hat has 4 sectors. Each of these sectors can be painted in one of the colors: blue or red. If a store contains all possible color options. How many options are there?

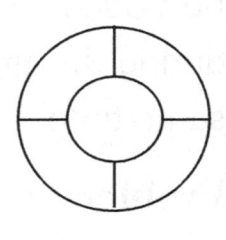

Problem 9. A cube has one of its 6 faces painted gray. By touching the paper with this face, the cube paints the paper gray. Sam rolls the cube on an edge in the checkered plane without touching the same cell twice. He draws a route to obtain each case shown in the figure. If initially the cube is upside down and the location of the cube is marked with a cross.

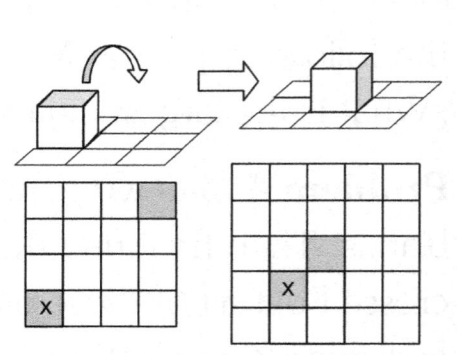

Problem 10. The game of "lying chips": Players take turns putting white or black chips on a board. If a player places a white chip, he must tell the truth; and if he puts a black chip, he must lie. There is 1 chip on the board. Petya put another chip on and said, "There are now more black chips on the board than white". What color is the first chip?

Problem 11. Two runners run one after the other at the same speed of 150 m/min, being at a distance of 300 m from each other. On the way, they came across a mountain. When going uphill, they reduced their speed by 50 m/min, and then on the descent they increased it by 100 m/min and then returned to their original speed. What is the maximum distance reached between the runners?

Problem 12. There is a chain of 13 links (each with a mass of 1 g), numbered in order: 1, 2, 3, ..., 13. Which link should be disconnected so that with the help of the parts obtained (including the free link) and a two-plate scale, can any mass of 1, 2, 3, ..., 13 g be measured with a single weighing? The parts obtained from the chain can be placed on both plates. After specifying the selected link, also indicate how the required measurements are obtained.

Problem 13. My grandmother has containers with a capacity of 7, 8 and 20 liters. The two smaller containers are filled to the brim with fruit juice and the larger one is empty. Can the fruit juice be divided equally in the three containers? If there are no additional devices, and there are also no graduation marks on the containers.

Problem 14. Gosha considers a month "successful" if it contains exactly 4 Mondays and 4 Tuesdays. Gosha once said: "By the way, this month is successful, last month was successful too, and next month will be successful too". In what month could Gosha say that?

Problem 15. In soccer, 3 points are awarded for a win, 1 point for a tie, and 0 points for a loss. In a 5-team soccer tournament, they play all against all (each played 1 time against the others). The "Meteor" team obtained 4 points; however, during the tournament this one scored 5 goals and received only 2 goals. Determine the scores for all the games played by Meteor.

Problems – Level 5

ANSWER SHEET – EXAM 9

Problem 1	Answer
Problem 2	Answer
Problem 3	Answer
Problem 4	Answer

Problem 5	Answer

Problem 6	Answer

Problem 7	Answer

Problem 8	Answer

Problems – Level 5

Problem 9	Answer

Problem 10	Answer

Problem 11	Answer

Problem 12	Answer

Mathematical Olympiads for Elementary School – Problem Book

Problem 13	Answer
Problem 14	**Answer**
Problem 15	**Answer**

Problems – Level 5

Exam 10

Problem 1. Write the largest number with different digits, where the adjacent digits differ by at least 2.

Problem 2. The monkeys Anfisa, Dusya and Musa were given bananas, no more than 10 in total. Anfisa gave 1 banana to Dusya and 2 bananas to Musa, after which they all had the same amount. How many bananas could Anfisa have initially? (List all options)

Problem 3. Cross out two cells to get the correct equality. The crossed-out cell does not participate in the calculations and could be anywhere, even between the digits of a number ("/" is the division sign).

| 4 | 8 | / | 1 | 2 | + | 3 | = | 5 | + | 7 | 2 | / | 1 | 2 |

Problem 4. Lesha and Kostya bought three pizzas, and received four pizzas of 28 cm, 18 cm, 16 cm and 6 cm in diameter, respectively - the last one as a gift for the order. Lesha ate the largest and the smallest, and Kostya ate the two medium ones. Who ate the most pizza?

Problem 5. At New Year's Eve, Santa Claus wants to make a magic square such that the sums of the numbers in the columns, the rows, and the two largest diagonals are equal. Put the missing numbers in the empty cells of the square.

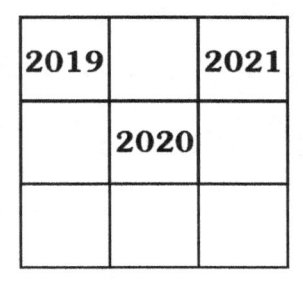

2019		2021
	2020	

Problem 6. In the city of Polossity, there are strange markings on every road: the first strip is 1 m long, then there is a 1 m gap, then a 2 m long strip, again a 1 m gap, then a strip of 3 m, a gap and so on until the road ends. If in the

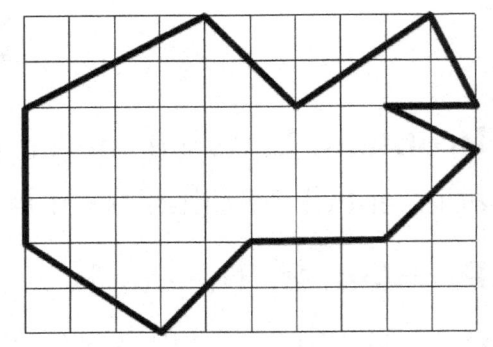

end there is not enough length of the road, then the strip just breaks. If the markings were made along a 2020 m long road. A) How many strips are there along the road? B) What is the length of the last strip?

Problem 7. Mark two points on the closed polyline drawn on the grid lines so that they divide the polyline into two parts of equal length.

Problem 8. To enter a dungeon, Harry Potter made an amulet out of identical magic cubes, as shown in the figure. But the insidious Draco Malfoy made six through holes with his wand, each of which went through exactly six cubes (parallel to the edges of the small cubes). A) How many small magic cubes are left intact if there are no gaps inside the amulet? B) How many small cubes were drilled three times?

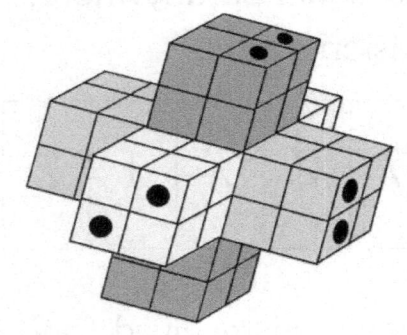

Problem 9. In the "rainbow" factory, a set of 15 plastic digits in 15 different colors is required to mount a wall clock, as shown in the figure. How many different dials can be made at this factory, if all sets of digits are exactly the same?

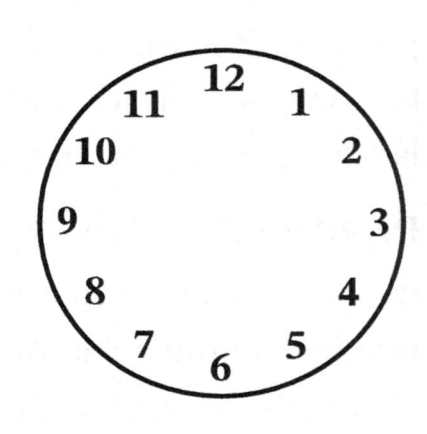

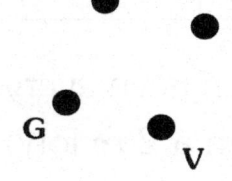

Problem 10. There were four people in a meeting: Alyosha, Borya, Vasya and Gosha. Alyosha said, "I have no friends here". Borya replied: "I have exactly 1 friend". Vasya in turn added: "I have exactly 2 friends". Gosha said: "I have 3 friends!". It turned out that everyone with an even

number of friends told the truth, and those with an odd number of friends lied. Draw a segment that connects two people who are friends.

Problem 11. The pistons of a trumpet were labeled with the letters A, B, C, D. One piston of the trumpet was pressed once, another was pressed twice, and the rest were pressed three times. How many times was piston A of the trumpet pressed, if piston B was not pressed three times, pistons A and C were pressed a different number of times, and pistons C and D were also pressed a different number of times?

Problem 12. Harry has 1 *ml* of a substance A in a 2 *ml* test tube, 4 *ml* of a substance B in a 4 *ml* test tube, 7 *ml* of a substance C in a 7 *ml* test tube. It is known that, if he mixes two different substances in equal proportions, he gets the third one. If the proportions are uneven, there will be a bang. Harry needs to get 4 *ml* of each substance. How can he do this, if he also has an empty 2 *ml* test tube? All test tubes are magic: the substance can be poured completely, and no trace will remain. There are no graduation marks on the test tubes.

Problem 13. In a square with side 4, another square with vertices at the midpoints of its sides is drawn. Likewise, the midpoints of one of the sides of each square are joined by a line segment (see figure). Find the area of the square whose side is equal to that line segment.

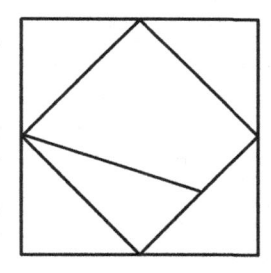

Problem 14. Kolya placed several colored cubes in a row on the table: white, blue, red, green, and black. It turned out that for any two colors there are a pair of cubes of these colors that are next to each other. That is, there are white and red cubes next to each other, green and white cubes next to each other, as well as for any pair of colors. What is the minimum number of cubes that can be on the table?

Problem 15. Gentlemen and liars live on the island of Romba (gentlemen only tell the truth, liars always lie). Exactly 1 person lives in each triangular area. In addition, areas that share one side are considered neighboring. In the morning, each of them said: "Among my neighbors there is only 1 gentleman". What is the largest number of gentlemen that can live on the island of Romba?

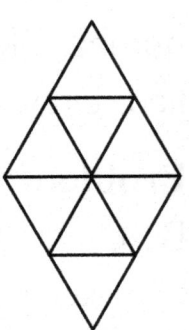

ANSWER SHEET – EXAM 10

Problem 1	Answer

Problem 2	Answer

Problem 3	Answer

Problem 4	Answer

Problem 5	Answer

Problem 6	Answer

Problem 7	Answer

Problem 8	Answer

Problems – Level 5

Problem 9	Answer

Problem 10	Answer

Problem 11	Answer

Problem 12	Answer

Mathematical Olympiads for Elementary School – *Problem Book*

Problem 13	Answer
Problem 14	Answer
Problem 15	Answer

Answers: Level 5

Answers – Level 5

Exam 1

1. 8 kg.
2. a) Sveta; b) 1 candy.
3. There are 7 roses.
4. XI – IV = VII
5. 977.
6. 4 pieces of paper.
7. "B-R" – blue and yellow, "B-Y" - red and yellow, "Y-R" – red and blue.
8. a) 15 cubes; b) 18 drops of gle.
9. An option is 2011 × (201 + 1).
10. 5 dots.
11. 12 large buckets.
12. 1° head - blows poisonous smoke; 2° head - turns to ashes; 3° head - shoots fire.
13. 4 cm^2.
14. Both options are possible.
15. Yes, it is possible. For example: 0, 1, 2, 3, 6.

Exam 2

1. 10 kg.
2. 120122012.
3. 80 cm.
4. 10 sheets.
5. An example is shown below:

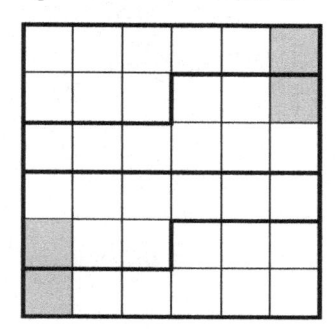

6. 88.
7. It was hot, snowing and raining.
8. 1 or 0.
9. At 3:00 pm.
10. 100 km.
11. Yes, it is possible. For example: 2012–2008–1992–2001–2003–2009–2027–2013–2011.
12. Stop B.
13. They are equal.
14. 6 boys.
15. The door C or F.

Answers – Level 5

Exam 3

1. The letter M.
2. An example is shown in the figure:

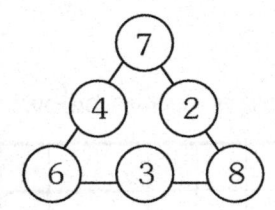

3. 10.
4. 600 m.
5. 61 cards.
6. An example is shown below:

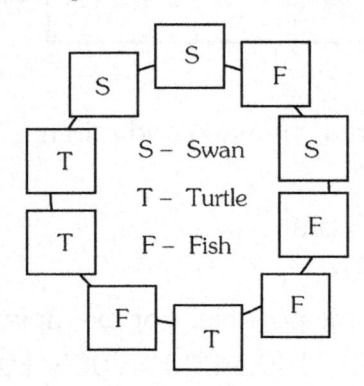

S – Swan
T – Turtle
F – Fish

7. 15.
8. On the first bed 19 and on the second bed 14.
9. 46 lamps.
10. The red petal.
11. An example is shown below:

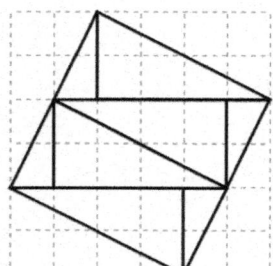

12. The second sage. **13.** No way.
14. 30 sales. **15.** Yes, it is true.

Exam 4

1. The number thought was 10.
2. 5 candies.
3. An example is shown in the figure:

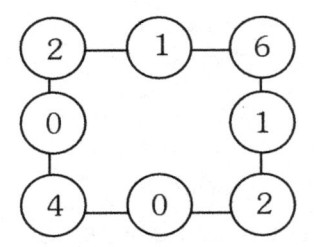

4. Petrovich.
5. The maximum is 24, and the minimum is 7.
6. 33 days.
7. In clockwise.
8. An example is $2 \times 6\,0\,1\,2\,0\,1 \times 4$.
9. The first is a liar, the second tells the truth, the third is a liar.
10. An example is shown here:

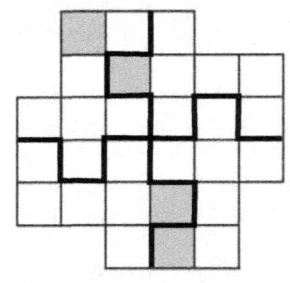

11. They live on opposite sides.
14. Tolya is right.
15. 6 meters.

Exam 5

1. 102346.
2. They planted the next trees: 4 pines, 3 spruces and 1 fir.
3. 16 times.
4. a) five or six brothers; b) 11 children.
5. 6 days.
6. a) 24 edges; b) 12 vertices; c) 14 faces.
7. 6 girls.
8. Prostofilin from Zmeyovka.
9. 793.
10. An example is shown below:

1	2	3	4	5	1	2	3
3	4	5	1	2	3	4	5
5	1	2	3	4	5	1	2
2	3	4	5	1	2	3	4
4	5	1	2	3	4	5	1
1	2	3	4	5	1	2	3
3	4	5	1	2	3	4	5
5	1	2	3	4	5	1	2

11. 10 seconds.
13. 510 *m*.
15. On 3 faces.

Exam 6

1. 10236.
2. An example is: 3+1+0-1+2-0+1=6.
3. 4 *cm*.
4. See the example in the next figure:

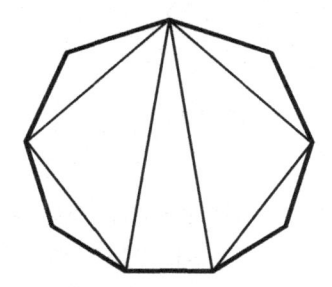

5. 10 students.
6. The maximum is 16 days and the minimum is 12 days.
7. Ocean is in the shaded circle:

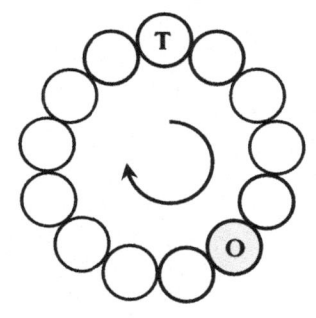

8. The white shape is as follows:

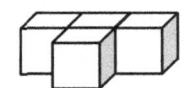

9. The maximum number of houses is 11, and the minimum is 2.
10. Masha, Pasha and Kolya.
11. It is not possible.
12. They are the same.
13. Day 1009.
14. No way.
15. Kolya-1, Pasha-2, Lesha-3, Sasha-4.

Answers – Level 5

Exam 7

1. It ends in zero (0).
2. Yegor has 18 and Artyom has 27.
3. 108 cm^2.
4. 2 boys.
5. 3488, 3489 and 3490.
6. 3 ways.
7. a) 7 min; b) 15 min.
8. 6 pieces.
9. Olya is in 5th grade, Vasya is in 7th grade, Masha is in 4th grade and Petya is in 6th grade.
10. 19 operations.
11. 1 sweet.
12. No, It could not.
13. 9 km.
14. February 13.
15. No, It could not.

Exam 8

1. An option is: 2017 + (7+ 1+ 0+ 2) : (2+ 0 + 1+ 7) = 2018.
2. The tunnel is 205 m long.
3. 8 numbers.
4. 15:30 or 3:30.
5. A) 13 dominoes, B) the last one is 0-6.
6. 24 in.
7. The possible options are marked in the figure:

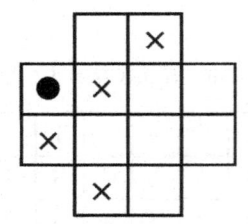

8. 12 dragons and 15 chuchundras.
9. An example is shown in the figure:

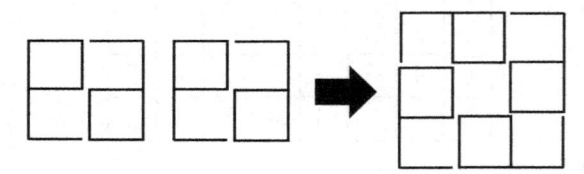

10. 3 liars.
12. The areas could be 30, 36 or 60 cm^2.
14. No, they cannot.
15. No way.

Answers – Level 5

Exam 9

1. 7210×921 or 9210×721.
2. 2 times.
3. The digit is 6.
4. A) 6; B) 0.
5. An example is shown in the figure:

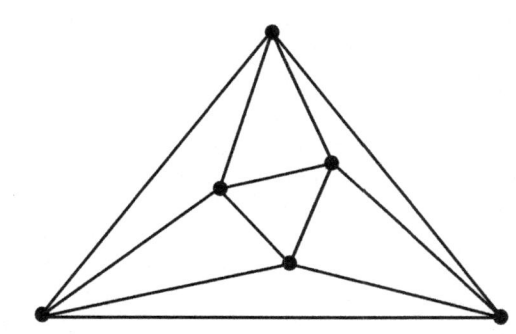

6. 3 cm^2.
7. January 12 at 12:00 noon.
8. There are 6 varieties.
9. The routes to follow are shown here:

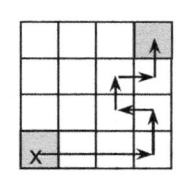

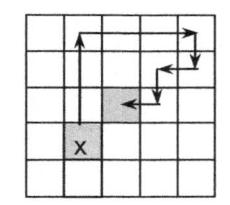

10. The first chip is white.
11. 400 m.
12. The fourth or tenth link.
13. No, it cannot.
14. In march.
15. The possible scores are: 0:1; 0:1; 5:0; 0:0.

Exam 10

1. 9758642031.
2. 5 or 6 bananas.
3. For example: 48/12 + 3 = 5 + 2/1 or 8/1 + 3 = 5 + 72/12.
4. Lesha.
5. An example is shown in the figure:

2019	2020	2021
2022	2020	2018
2019	2020	2021

6. A) 63 strips; B) 5 m long.
7. An example is shown below:

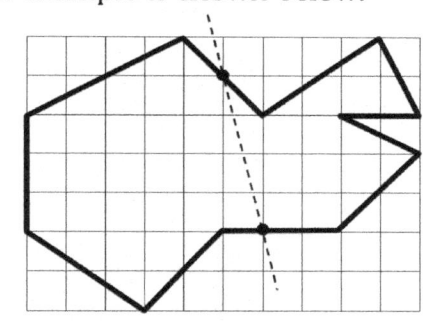

8. A) 25 cubes; B) 1 cube.
9. 480 dials.
10. Draw the segments as follows:

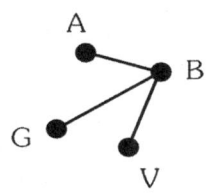

11. 3 times.
13. 10 cm^2.
14. 11 cubes.
15. 6 gentlemen.

Answers – Level 5

Made in United States
North Haven, CT
23 November 2022